Rocks,
Minerals
& Fossils
of the
World

Rocks, Minerals
& Fossils
of the World

CHRIS PELLANT

Photographs by Roger Phillips
& Chris Pellant
Layout by Jill Bryan

Little, Brown and Company
Boston   Toronto   London

Acknowledgements

It would have been impossible to produce such a complex book without the help of many people. I must especially thank Professor Gilbert Kelling and his staff in the Department of Geology at the University of Keele, UK, for the loan of many specimens used for photography, and for much other help and advice. Hugh Torrens has been of great assistance with reference material, the loan of specimens and checking information; his depth and breadth of knowledge never cease to amaze me! The specimens borrowed from Keele were catalogued and packed by Andy Lawrence, who carried out this daunting task with great efficiency. George Rowbotham and Phil Lane have allowed me access to their collections and David Stevenson helped with the section on echinoderms.

Other people and organizations who have helped in a variety of ways, including the loan of specimens for photography are: Bob Owens at the National Museum of Wales in Cardiff, Ian Rolfe and Bob Reekie at the Royal Museum of Scotland in Edinburgh, Jim Nunney at Leeds City Museum, Don Steward at the City Museum in Stoke-on-Trent, Don Mason at the Dorman Museum in Middlesbrough, and the Whitby Museum. Ken Sedman of Cleveland County Leisure Services Geology Department provided many fine specimens for photography and gave valuable advice and access to reference material.

David Curry, Arthur Forsey, John Fraser, Gordon Leedale, John and Hilda MacPherson, Trevor Morse, Jeff Mullroy, Bruce Stevenson, Pat Tye, Lynn Wall and Sid Weatherill all helped in various ways. I also had photographs from Martyn Rix (pages 50, 53), Jill Bryan (pages 37, 99), David Curry (pages 29, 143), Pat Tye (pages 15, 19), Arthur Forsey (page 36) and Bob Reekie (page 33).

Throughout the project my wife, Helen, has been a constant source of support and encouragement. She has corrected proofs and compiled the index with meticulous attention to detail. As well as thanking her I must also thank my three children, Daniel, Adam and Emily, for not making too much noise over the last two years!

Text and photographs copyright © 1990 by Chris Pellant.
Studio photographs copyright © 1990 by Roger Phillips

First edition

Library of Congress Catalog Card Number 89–63627
Library of Congress Cataloging-in-Publication Information is available

10 9 8 7 6 5 4 3 2 1

Published simultaneously in Canada
by Little, Brown & Company (Canada) Limited

Printed in Singapore

3 4 0 1 5 4

Contents

Introduction

Rocks, minerals and fossils are the raw material of the science of geology. They may be either the basis of high level research, or the eminently collectable items which are prized by amateurs and professionals alike.

Identification of specimens comes with practice. The more specimens you investigate the better you will become at understanding and classifying them. It is important to adopt a logical approach, looking at each aspect of the specimen in turn. It will not always be possible to match a specimen with an illustration, but the photographs and descriptions herein will, it is hoped, provide a fair level of identification if not absolute naming.

The **minerals** are arranged in the conventional manner starting with native elements and continuing with sulphides, oxides and so on. The specimens chosen include some perfect crystals but are mainly typical material rather than exquisite museum pieces. It is hoped that this will make it easier for you to match a specimen to a photograph. Each mineral specimen is accompanied by a full description of its properties and some indication of its distribution along with its more usual modes of occurrence.

The **rocks** are divided into three sections: igneous, metamorphic and sedimentary. In this section of the book there are a number of 'field' photographs, for it is in the field that geology comes alive and where the relationships between strata and their contents are best understood.

The **fossils** are the most difficult group to identify. They are arranged in biological groups and within each group the genera are stratigraphically organized as near as is possible. It is not easy to identify fossils below generic level from photographs and naming has been left at this level in nearly every case. This is, nevertheless, a good level to reach. For more detailed classification

one of the standard reference works will have to be consulted, or you may need to visit a local museum or university department. I have given an indication of geographical distribution. However, it should be borne in mind that the distribution of species is constantly being modified by fieldwork and research. The expression 'worldwide' is sometimes used and this obviously means only where favourable strata occur. The geological range of each genus is given with each description. Microfossils have not been included because they require certain specialized techniques of study and are numerous enough to fill a book on their own.

When collecting specimens be sure to place each item in a separate container with the name of the locality carefully recorded. The importance of this location of material cannot be over-emphasized, and is of equal importance to the name of the specimen; indeed, the location will help when trying to name the material. Try to make a catalogue of your collection and if it becomes extensive then lodge a copy with your local museum, as you may have material which, if properly curated, is of scientific

Halysites

The Painted Desert, Arizona, USA

Hematite

interest. It is a shame to think that much of our geological heritage is hidden away in attics and garages! When out collecting be sparing with the use of your geological hammer. Use it for breaking up loose blocks, rather than for quarrying the exposure, and collect in moderation. A camera is a valuable piece of field equipment, and today with good cameras available at modest costs even the most distant locations can be recorded on film, together with the specimens they contain.

On the main plates the black circle is one centimetre in diameter.

This book has been designed for anyone who has an interest in the materials of the Earth. It is hoped that in helping you to understand and identify specimens you will develop a respect for the Earth and an awareness of the need to conserve the geological record that is preserved in the rocks of the crust.

The people and organizations who have lent specimens are acknowledged elsewhere but the captions for each specimen have the following key letters in brackets to indicate the origin of that specimen: AF Arthur Forsey; BS Bruce Stevenson; C Cleveland County Leisure Services; CP Chris Pellant; D Dorman Museum, Middlesbrough; HST Hugh Torrens; JHF John Fraser; JM Jeff Mullroy; JMc John and Hilda MacPherson; K Keele University; L Leeds City Museum; LW Lynn Wall; RMS Royal Museum of Scotland; S City Museum, Stoke-on-Trent; SW Sid Weatherill; TM Trevor Morse; W National Museum of Wales; Wh Whitby Museum.

The Painted Desert's Triassic sandstones and mudstones of Arizona, USA, owe their striking colours to a variety of minerals which have cemented the grains of sediment. These are mainly iron compounds, but manganese, copper and uranium minerals are also present.

The Photographs

My field photographs and close-ups of most of the individual specimens were taken on 25 ASA or 64 ASA Kodachrome film in Olympus 35mm cameras. Lenses ranging in focal length from 28mm to 70mm were used for the location pictures. The mineral specimens were photographed with a 55mm macro lens with twin electronic flash. Close-ups of rock specimens were taken in a similar way, though some which were taken in the field relied on available light. In field situations a tripod is an essential piece of equipment since, even in poor light, the ability to use a slow shutter speed (geological materials tend not to move about!) allows greater control over depth of field. Indeed, soft, almost shadowless light is very good for photographs of many specimens. The pictures of fossils were taken on both daylight film, using lighting as described above, and on tungsten balanced Ektachrome 50 ASA film with photofloods.

Roger's studio photographs were taken on a Hasselblad with a planar 80mm lens and Kodak Ektachrome 64 ASA film.

Collecting minerals on old mine spoil heaps, North Yorkshire, UK

Glossary

AEOLIAN: A term usually applied to sediments, indicating that they are wind deposited.

ARENACEOUS: Sediments containing much sand.

AUTHIGENIC: The formation of minerals in place after or during deposition.

AUREOLE: The zone surrounding an igneous intrusion which is affected by heat, and is thus metamorphosed.

BASAL CLEAVAGE: Cleavage in a mineral which is parallel to the basal pinacoid.

BASIC: A term referring to an igneous rock with less than 10 per cent free quartz and a total silica content between 45 per cent and 55 per cent. Such rock consists mainly of plagioclase feldspars and pyroxene.

BENTHONIC: On the sea floor.

BIFURCATION: The splitting into two of, for example, the ribs on the shell of a mollusc.

BIO-MICRITE: Limestones consisting of organic fragments, or whole shells, in a matrix of calcite ooze.

BRACHIAL VALVE: The valve of a brachiopod which is generally the smaller of the two, and which contains the brachidium or lophophore.

CALCRETE: Gravels cemented by tufa.

CALICE: The hollowed depression in the top of a corallite in which the polyp is housed.

CEPHALON: The head-shield of a trilobite.

CILIATE: Fringed with small hair-like structures.

CIRRI: Prehensile branches on the stems of crinoids.

CLEAVAGE: In rocks this is the fine parting developed during low-grade regional metamorphism which allows slate to be split into thin sheets.

COLLOIDAL: The state between suspension and solution which allows ultra-fine particles not to sink through a fluid medium.

CONCHOIDAL: A curved or shell-like fracture which is characteristic of some minerals and rocks including quartz and obsidian.

CONCRETION: A mass of rock usually rounded in shape and a few inches in diameter, often occurring in clay or shale strata. Concretions can be made of ironstone or limestone, or may be chemically the same as the strata they are in. Fossils are often contained in concretions, being perfectly preserved in three-dimensional form.

CORALLITE: An individual coral unit. These may be solitary or joined in a colony.

CRYPTOCRYSTALLINE: A mass of minute crystals forming an aggregate, the individual parts of which can only be distinguished at very high magnification.

DIAGENESIS: The processes which turn soft sediment into consolidated rock. These occur near the surface of the Earth at low temperatures and pressures and include chemical and physical processes.

DRUSY: A drusy cavity is a hollow in a rock such as granite, or in a mineral vein, where well-formed crystals project.

EQUIGRANULAR: A rock texture where all the grains are the same size.

EUHEDRAL: Well-formed crystals.

EVOLUTE: Loose coiling of a cephalopod shell where the outer whorls (coils) do not overlap the inner ones much. All the whorls can thus be seen and the umbilicus is wide.

EXTRUSIVE: Volcanic rocks; molten rock which has cooled on the surface.

FOLIATION: A parallel orientation of flaky minerals, such as mica, in a rock.

GENAL SPINE: A spine in some trilobites which extends from the cephalon towards and along the side of the thorax.

GRAPHIC TEXTURE: An intergrowth of quartz and feldspar which occurs in some igneous rocks, especially granites. When seen on a smooth surface this gives the effect of runic or hieroglyphic writing.

HETEROCERCAL: A fish tail with the vertebral column in the upper lobe. This lobe is larger than the lower lobe.

HOMEOMORPHY: The state of two not necessarily related organisms exhibiting a very similar outward form.

HOPPER CRYSTALS: A crystal with faces which are hollowed, as in the case of cubes of halite.

HYPABYSSAL: Igneous rocks which have crystallized at shallow depth, often in relatively small bodies.

INTERSTITIAL: Between the crystals or grains of a rock.

INTRUSIVE: Igneous material formed among pre-existing rocks below the surface. Magma invades these rocks and thus intrudes them.

INVOLUTE: The opposite of evolute, when the whorls of a cephalopod shell are tightly coiled and overlap obscuring the inner whorls and producing a narrow umbilicus.

METASOMATISM: A change in a rock brought about by the introduction of a new material, often by fluids circulating in the crust.

METEORIC WATER: Water entering the rocks from above, including atmospheric precipitation and river waters.

NEMA: A thread-like extension at the end of a graptolite's stipe.

OPHITIC: An igneous texture in which lath-shaped plagioclase crystals are enclosed in pyroxene.

OROGENY: Mountain building. All the various processes which lead to the development of a mountain range. This will take tens of millions of years.

ORTHOCONE: An elongate, straight molluscan shell, as in some forms of nautiloid.

PEDICLE VALVE: The valve of a brachiopod which is the larger of the two and which has the pedicle opening.

PELITIC: Muddy or clayey sediment (= argillaceous).

PERIPROCT: The flexible plated region surrounding an echinoid's anus.

PERISTOME: The area of plates surrounding an echinoid's mouth.

PINACOID: A pair of parallel crystal faces.

PINNATE: A leaf with leaflets arranged on both sides of a central stem.

PINNULES: Crinoids' arms branch and the branches may have rows of smaller branchlets called pinnules.

PLANKTONIC: Near the sea surface, floating and moved by currents.

POIKILITIC: A rock texture where small crystals in an igneous rock are enclosed in a larger crystal.

PORPHYROBLAST: A crystal set in the matrix of a metamorphic rock, which is larger than the matrix material, for example euhedral garnet crystals in schist.

PYGIDIUM: A trilobite's tail segments.

PYROCLASTIC: Fragmental volcanic material including bombs, pumice and tuff.

PYROXENE: A common group of silicate minerals. The

Oolitic limestone

orthopyroxenes form orthorhombic crystals; the clinopyroxenes are monoclinic. There are also chemical differences.

SACCHAROIDAL: A sugary rock texture not uncommon in quartzites and marbles.

SEPTA: The partitions which separate chambers in a mollusc shell or in a coral.

SCHISTOSITY: The wavy fabric of a medium-grade metamorphic rock created by the alignment of minerals.

SCHILLERIZATION: A striking play of colour in certain minerals, especially plagioclase feldspars, due to the numerous minute, rod-like inclusions of iron ore.

SCORIACEOUS: Lava texture which contains hollow cavities.

SIGMOIDAL: 'S'-shaped.

SILICATE: The most abundant group of minerals in the Earth's crust including the important rock-forming feldspars, amphiboles and pyroxenes.

SIPHUNCLE: A thin tube running from the body chamber through each septum to the protoconch of a cephalopod.

SPICULES: The siliceous units of a sponge skeleton.

STOCK: An intrusive plutonic igneous body, smaller than a batholith and commonly more or less circular.

TELEOST: Bony fish of 'modern' type with skull bones covered by outer skin and with a symmetrical tail.

THECODONT: A group of Triassic bipedal reptiles from which dinosaurs and crocodiles developed.

THRUST: A fault with a very low angled plane where a mass of older rock has moved over younger strata.

TWINNING: When two parts of a crystal have a different orientation but share some crystallographic direction or plane.

ULTRA-BASIC: The group of igneous rocks which generally has less than 45 per cent total silica. The ultra-basic rocks contain much ferro-magnesian material and are often almost mono-mineralic. For example the ultra-basic rock dunite is made of olivine.

UMBILICUS: The centre of a coiled molluscan shell.

VESICULAR: A texture in igneous rocks, usually lavas, consisting of a matrix filled with gas-bubble cavities.

WHORL: A single coil of a molluscan shell.

ZEOLITES: Silicate minerals which have water of crystallization in their structure. They often occur in amygdales and include harmatome, natrolite and chabazite.

Geological time-scale

ERAS	PERIODS			EONS
	Quaternary	Recent	0.01	
Cenozoic		Pleistocene		
			— 2	
		Pliocene		
		Miocene	5	
	Tertiary	Oligocene	22.5	
		Eocene	37.5	
		Palaeocene	53.5	
			— 65	
	Cretaceous			
			— 136	
Mesozoic	Jurassic			
			— 190	
	Triassic			
			— 225	
	Permian			
			— 280	Phanerozoic
Upper Palaeozoic	Carboniferous			
			— 345	
	Devonian			
			— 395	
	Silurian			
			— 440	
Lower Palaeozoic	Ordovician			
			— 500	
	Cambrian			
			— 600	
				Proterozoic
Pre-Cambrian			2500	
				Archaean
			— 4600	

Calamites

9

The 'Cheesewring' tor, Bodmin Moor, Cornwall, UK

Igneous Rocks

Igneous rocks are formed by the crystallization of magma or lava, a complex silicate melt which occurs in a variety of environments in and on the Earth's crust. The way magma is injected into the crust determines many of the features of the resulting solid rock, especially the size and relationship of the crystals of which it is formed, that is the rock's texture. Batholiths and plutons are enormous bodies of magma which are often intrusive, but may also partly melt their way into the crust. Because of their size and depth, the magma cools very slowly, possibly taking tens of millions of years from initial injection to freezing. The rocks thus formed have large crystals due to the slow cooling, and are coarse grained (over 5mm long). Granite is one of the commonest rocks in these bodies, which are often emplaced in the roots of mountain chains. In British Columbia and Peru the coastal batholiths are over 1600 km long and 190 km across; the batholith in south west England is small by comparison at 64 km by 32 km. Other large intrusions may be saucer shaped (lopoliths), being fed by a vertical pipe. Such a feature in South Africa measures 480 km by 320 km. A laccolith is similar but with a convex upper surface. These large intrusions contain rock made of coarse crystals, but in smaller bodies the magma can freeze more quickly and finer-grained rocks result. When magma wells up through weaknesses in the crust the resulting igneous intrusions are often sheet-like, being narrow but extensive. Discordant sheet intrusions, which cut through existing strata, are called dykes. These are very common and often vertical. Sills are sheet intrusions which follow existing structures and are thus concordant. Because these narrow intrusions (they may be only a few metres wide) cool quite rapidly they contain rocks with medium- to fine-grained crystals (5–0.5 mm and less than 0.5 mm). Dolerite and micro-granite are common rocks in such intrusions. Extrusive lava cools very rapidly, even faster than a small sill or dyke, and so the rocks resulting from volcanic activity will be fine-grained like basalt and andesite. If the volcano erupts into water, as is often the case, very rapid freezing may produce a glass without recognizable crystals.

Grain size is thus closely related to the speed of crystallization, but the detailed cooling history of the magma, or lava, will produce particular textures. An attractive rock texture which has large euhedral crystals (phenocrysts) set into a relatively finer matrix is the porphyritic texture developed by two definite stages in the cooling process. If the magma undergoes a slow cooling period deep in the crust large well-formed phenocrysts may develop within the liquid magma. If this magma is then carried higher into the crust, or even on to the surface, the rest of the magma will freeze rapidly producing a finer matrix around the phenocrysts.

Grain size is one of the two main criteria used for identifying and classifying igneous rocks. As a general guide the crystals in a coarse-grained rock like gabbro or granite can be easily seen with the naked eye. Those in medium-grained rocks, such as dolerite and micro-granite, are best examined with a hand lens, while in fine-grained lavas like rhyolite or basalt the grains cannot be easily distinguished even with a hand lens. When a microscope is used to examine igneous rock, and polarized light is passed through a very thin slice, it will reveal great detail of texture and chemistry. Certain minerals exhibit striking colours which can be a great help in identification, and it is amazing how a fairly dull-looking rock like dolerite becomes an astonishing mosaic of colours and shapes in thin section.

The other criterion used to identify an igneous rock is its chemistry and mineral composition. Magma may originate in the Earth's mantle, below the crust, many hundreds of miles below the surface. The composition of this rock is low in total silica and is said to be ultra-basic. (The term 'basic' means low in silica; acid rocks have a high silica content.) The ultra-basic rocks of the mantle are rich in iron and magnesium and are similar to the peridotite which occurs in the crust and can be brought from great depth by volcanic action. Even though magma from the mantle is of ultra-basic composition it can give rise to a variety of rocks through magmatic differentiation. Recent research has shown that even acidic rocks with over 65 per cent total silica can arise from this magma. Also, pre-existing rocks can be melted and assimilated to produce a magma of high silica content. However the magma is generated, only a relatively few minerals make up the igneous rocks. Feldspars, quartz, mica, amphiboles, pyroxenes and olivine are the commonest rock-forming silicates. They crystallize in a definite sequence as the magma or lava solidifies. At the highest temperatures the dense silicates of iron and magnesium form, olivine first followed by pyroxenes. These 'pyrogenetic' minerals contain no water in their molecular structure. As the temperature falls, other silicates like amphibole (including common hornblende), which does contain water, form, and finally biotite mica crystallizes. The minerals in this ferro-magnesium series do not grade into each other but form at definite temperature intervals.

While these dense silicates are forming, a family of pale, lower-density feldspars, the plagioclases, will also be crystallizing. At about the same temperature as olivine and pyroxene, calcium-rich plagioclase, anorthite, freezes. As the magma cools sodium replaces calcium in the plagioclase structure and this allows a continuous series of feldspars to develop with sodium-rich albite at the lower temperature end and a plagioclase with 50 per cent calcium and 50 per cent sodium midway between anorthite and albite. Towards the end of the cooling sequence, orthoclase, muscovite and quartz form. The sequence outlined here is idealized. Not all these minerals will occur in one rock. A magma may cool in such a way that only the high temperature minerals develop, or the temperature of the magma may be low when it forms thus only allowing minerals at the bottom end of the scale to crystallize. Basic magmas, generated in the mantle, are very high temperature melts and from these olivine, pyroxene and calcium plagioclase develop, forming rocks like basalt and gabbro. More acid magmas, possibly formed at higher levels, are of lower temperature and from them quartz, orthoclase and mica-rich rocks crystallize, producing granites.

This brief outline indicates the ways in which igneous rocks may develop their characteristic features. In order to identify a specimen, a logical approach is best. Examine the grain size and decide whether the rock is coarse-, medium- or fine-grained. Then consider the overall colour of the rock. If it is pale coloured it will tend to be acidic; darker rocks are more basic in their mineralogy. The minerals can be examined using a hand lens. There will often be only about four minerals to identify and the relative abundance of these will indicate whether the rock is acid, intermediate, basic or ultra-basic in composition. The presence or absence of quartz, which is easy to identify by its great hardness and grey vitreous appearance, helps to pin the rock down. The common rocks are relatively easy to identify. Granite can be identified by its coarse grain size, pale colour and abundance of free quartz. Dolerites and basalts are dark and heavy with small laths of feldspar and pyroxene visible in the dolerite, especially with a lens. The ultra-basic rocks may be less common but because they are often made of only one or two minerals and lack common quartz, they are not too difficult to identify. As with many types of identification, experience counts for a great deal; the more rocks you see the better you will become!

The 'Cheesewring' tor on Bodmin Moor, Cornwall, UK. When granite is weathered on the Earth's surface, landforms called tors develop. These often have strange shapes and are given local names. Here the granite of the 'Cheesewring' tor seems to have a layered structure. The horizontal partings are weathered joints which, along with vertical ones, are thought to develop when the magma

consolidates and the once deep-seated granite is relieved of the weight of overburden.

Granite seen at a magnification of × 3, is an interlocking mosaic of crystals. The pink are orthoclase feldspar, the grey, glassy crystals are quartz and the small black specks are biotite mica. There are also a few white crystals of plagioclase. This granite has an equigranular texture, all the crystals being much the same size. The specimen is from South Africa. (CP)

Basalt is very different from granite in a number of ways, these two igneous rocks being at opposite ends of the classification scheme. This picture is taken at the same magnification as the one of granite and shows the very fine-grained nature of this extrusive, rapidly cooled rock, and the typical vesicular texture, with many gas-bubble cavities. The specimen is from Hawaii. (CP)

Basalt

Granite

This is a very common coarse-grained igneous rock of acid composition. The texture is often equigranular, but phenocrysts of feldspar may give a characteristic porphyritic appearance. Such crystals can grow to well over 5 cm in length. Granites contain at least 10 per cent free quartz, usually between 20 per cent and 40 per cent. Both orthoclase and plagioclase feldspars are present; the latter tends to be sodium rich. When orthoclase and plagioclase are present in equal amounts, and the plagioclase has a more calcium rich chemistry, the rock is called adamellite; when the plagioclase becomes dominant, granite grades into granodiorite. With a decrease in the quartz content granite becomes syenite. The presence of much quartz and feldspar gives granite a pale colour, though the orthoclase is often pink. Mica is an essential constituent; both pale muscovite and brown or black biotite can occur, either together or separately. Hornblende is a common ferromagnesian constituent. The presence of biotite and hornblende gives the rock a speckled aspect. Accessory minerals found in granite include sphene, magnetite and apatite. Granites can vary a great deal in their detailed chemical composition and this leads to many varieties each with a distinctive appearance, and to special textural features like graphic or runic texture where quartz 'characters' are embedded in microcline. Very coarse-grained granites with all grains over 3 cm long are called pegmatites. Granites occur in a wide variety of intrusive bodies, typically in very large plutonic batholiths. The rock certainly has a magmatic origin but some granites may arise from granitization associated with fluids active at great depths in the crust.

Xenolith in granite The dark ovoid mass, which is about 10 cm long, is a fragment of country rock which has been surrounded by granitic magma and partially assimilated (a xenolith). It has retained something of its original character; it was a block of andesitic lava. The granite is porphyritic but the pink orthoclase is in two sizes, phenocrysts and groundmass crystals.

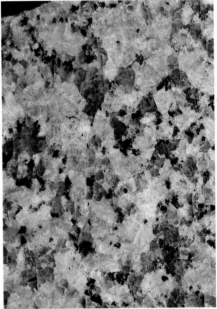

Granite seen at a magnification of × 3

Xenolith in granite

Xenolith

Porphyritic adamellite

Biotite granite

Pink granite

Muscovite-biotite granite

Biotite-muscovite granite

Muscovite-biotite granite from northern Spain

Biotite granite from SW England

Muscovite granite

Muscovite granite from the Isle of Man

Xenolith Fragments of country rock which are incorporated in magma become partially or wholly altered. This specimen was originally volcanic tuff and it has now been partially granitized by the surrounding magma. Pink orthoclase feldspar crystals have developed in the xenolith. (CP)

Biotite granite A specimen from Germany which is very pale in colour due to much white orthoclase and plagioclase and grey quartz. Small black crystals of biotite mica give the rock a faintly speckled appearance. (CP)

Biotite-muscovite granite A specimen with much euhedral white orthoclase and albite (plagioclase). There is over 30 per cent quartz which is grey in colour, and biotite mica is in excess of white muscovite. Some of the orthoclase crystals are up to 8 mm long. Apatite is an accessory mineral. The specimen is from a borehole in north-east England. (C)

Pink granite This specimen from South Africa has an equigranular texture with much pink orthoclase, grey quartz and white plagioclase. Small amounts of both muscovite and biotite mica are present. The pink orthoclase is well in excess of the plagioclase. (CP)

Muscovite-biotite granite A very pale granite from Georgia, USA, which has euhedral white orthoclase and albite, grey anhedral quartz and both micas. Muscovite is in excess of biotite. (C)

Biotite granite from south-west England A specimen with a black and white appearance containing milky-white orthoclase and pale plagioclase. There is much anhedral grey coloured quartz and some 10 per cent black biotite. Though the texture is predominantly equigranular, some of the orthoclase crystals are much larger than the plagioclase and quartz. (K)

Muscovite-biotite granite from northern Spain A greyish granite with an equigranular texture. Orthoclase is well in excess of plagioclase (albite) and there is more muscovite than biotite. Grey quartz accounts for about 30 per cent of the rock. (CP)

Muscovite granite A specimen from north-west France which is very pale and contains up to 12 per cent muscovite mica. The matrix consists of euhedral crystals of orthoclase and plagioclase and anhedral quartz. (C)

Muscovite granite from the Isle of Man, UK. This specimen consists of off-white orthoclase, white plagioclase, grey quartz, and muscovite up to 10 per cent. (L)

Granites and Pegmatites

Pegmatites are very coarse-grained igneous rocks with a grain size over 3 cm. They contain some of the largest crystals found in the Earth's crust and because they often form in the final stages of magmatic crystallization, when fluids and residual solutions containing rare elements

Granite seen at a magnification of × 16

Granite seen at a magnification of × 16. This thin section of granite from Rubishlaw, Aberdeen (Scotland), shows much pale and grey quartz, with feldspar readily identified by its 'striped' appearance. The large brown crystal is biotite mica. (CP)

Porphyritic adamellite has large phenocrysts of orthoclase feldspar giving the rock an appearance which has been exploited for ornamental stone. These euhedral crystals are set in a matrix of grey quartz, white oligoclase and black biotite. The percentages of orthoclase and oligoclase are equal. The specimen is from Cumbria, UK. (CP)

are abundant, so pegmatites are rich in uncommon minerals. These include tungsten and radioactive minerals. Pegmatites are closely related to granites and their mineral content is very similar, with alkali feldspar dominant, quartz and micas also being in relatively high proportions. All these minerals grow to large size as do tourmaline, beryl, sphene, fluorite, rutile and a wide range of other accessories. Graphic texture is a common feature but the amazing coarseness of the grains is such that crystals up to 20 metres long have been found. Pegmatites often form in dykes and veins near the margins of the plutonic masses. These dykes are usually of no great geographic range and are not as consistently sheet-like as other dykes. Though the rock name pegmatite is used for exceptionally coarse-grained granitic material, the term pegmatitic can be applied to any very coarse variety of other igneous rocks.

Perthitic granite   A coarse-grained rock with much grey vitreous quartz. Muscovite mica is dominant over biotite and there is a mingling of potassium and sodium feldspar (perthite). The rock is typically pale coloured and of low density. (CP)

Pegmatite   Very coarse crystals mainly of orthoclase feldspar with black biotite forming large tabular masses. There is some grey anhedral quartz. This specimen is from Sutherland in north-west Scotland. (CP)

Luxullianite   Tourmaline is a common accessory in many granites and in this rock pink orthoclase has much altered margins surrounded by black tourmaline and some grey quartz. The specimen is from Carnmanellis in Cornwall, UK. (K)

Coarse-grained pegmatite with large plates of muscovite mica among grey quartz and white feldspar. (CP)

Pegmatite   Granites often have veins and drusy cavities which contain fine crystals of quartz and other minerals. In this specimen veins of white quartz alternate with patches of muscovite and orthoclase-rich granite. (CP)

Hornblende granite   The dark minerals in this specimen from Sweden are hornblende and biotite. Hornblende is not uncommon in granites and gives such rocks a dark speckled appearance. The pale minerals in this specimen are grey quartz and white orthoclase and oligoclase. (C)

Graphic granite   It is not uncommon for granites to have a graphic or runic texture which occurs when small angular particles of quartz are set in a large alkali feldspar groundmass. The feldspar is often microcline. This texture is well-known in pegmatites. Graphic granites are very pale-coloured rocks. (K)

Pegmatite intrusions often occur as sheets cutting through country rock. Here a feldspar-rich pegmatite about 60 cm thick intrudes dark ultra-basic Pre-Cambrian rocks. Higher up the cliff is a small vein of similar pegmatite. The exposure is on the coast of north-west Scotland.

Perthitic granite

Pegmatite

Luxullianite

Pegmatite

Coarse-grained pegmatite

Hornblende granite

Graphic granite

Pegmatite intrusions

Microgranite

Microgranite from Norway

Greisen

Granodiorite

Porphyritic microgranite

Microgranite

Quartz porphyry

Granodiorite

Frothy ignimbrite flows

Microgranites, Greisen and Granodiorite

Microgranite A medium-grained rock with much pink orthoclase feldspar forming a groundmass in which there are phenocrysts of biotite and larger 'plates' of silvery muscovite. (CP)

Microgranite from Norway containing equigranular crystals of medium-grained grey quartz, white feldspar and black biotite mica. (CP)

Greisen is produced by the effect of fluids and gases on already formed granite. A rock rich in muscovite and quartz and yellowish-coloured gilbertite mica is produced. Such rocks are often marginal to granite bodies or vein rocks in fissures in the granite. The most common volatiles which alter granites in this way are fluorine, hydrofluoric acid and boron fluorides. (CP)

Granodiorite is a pale to medium-dark coloured rock of coarse to medium grain size. There can be up to 20 to 40 per cent free quartz but the total silica content is lower than in granite; there is also a higher percentage of calcium and magnesium. The typical plagioclase feldspars are andesine and oligoclase, with orthoclase, biotite and hornblende giving the rock a very similar appearance to granite. Two specimens are shown; the paler one is from southern California, USA, the darker one is from Brittany, France. (L)

Porphyritic microgranite Dark crystals of amphibole (riebeckite) in a pale groundmass of alkali feldspar and quartz. When seen in thin

Granodiorite seen at a magnification × 16

section this rock shows poikilitic texture, small crystals embedded in a larger one. (L)

Microgranite A medium-grained equigranular rock with pale pink orthoclase, grey quartz and both biotite and muscovite mica. This specimen is from New Hampshire, USA. (L)

Quartz porphyry This specimen has a medium-grained matrix of orthoclase, quartz, hornblende and biotite. The phenocrysts are of euhedral feldspar and some quartz and their presence suggests a two-stage cooling history. Such rocks form in minor intrusions. (K)

Granodiorite seen at a magnification of x 16. The brightly coloured crystals are mainly hornblende, but there are some small brown biotite grains. The large grey crystal to the right

of centre is plagioclase showing twinning. The other grey materials are quartz and orthoclase. The specimen is from Leicestershire, UK.

Granophyre/Acid Lavas

Frothy ignimbrite flows In this cliff section in Costa Rica two distinct layers are visible, an upper grey mass and a lower yellowish section. The grey is an ignimbrite flow produced by shallow level degassing of silica-rich viscous magma. Frothy ignimbrite flows are associated with violent eruptions but if the degassing takes place at some depth, a clastic air-fall deposit is produced, the lower yellowish layer in the picture.

Pitchstone

Granophyre

Rhyolite tuff

Obsidian

Granophyre

Obsidian

Porphyritic pitchstone

Banded rhyolite

Pitchstone is a rather shiny rock with the lustre of tar or pitch. The rock has cooled very rapidly as lava and so a glassy, almost crystal-less, texture has developed. This example from the Isle of Arran, Scotland, is of acidic composition but pitchstones can vary to nearly basic composition. This is a rock of minor intrusions and lava flows. (C)

Granophyre is a pale-coloured rock often with a pinkish tinge and medium grain size. The composition is granitic with more than 10 per cent quartz, much orthoclase, albite to oligoclase as the plagioclase feldspars, mica and some amphibole. In granophyres the quartz and feldspars tend to display a texture of intergrowth, termed granophyric. This is a finer-grained version of graphic texture. Granophyre is a rock from the margins of large intrusions and hypabyssal bodies; it is closely allied to porphyritic microgranites. Two specimens are shown. (CP)

Rhyolite tuff is a fragmental volcanic rock made of particles of consolidated rhyolite lava welded together in a siliceous matrix. Tuff generally contains fragments of small size (ash) which have been blown from a violent volcanic eruption. Such material may settle in water and form neat strata; finer-grained material settles more slowly and graded beds may then develop. (K)

Obsidian is an extremely shiny glass with a typical conchoidal fracture which produces very sharp edges when broken. It is usually black or dark green coloured. The composition is similar to rhyolite and on rare occasions it contains phenocrysts. (K)

Banded rhyolite is chemically similar to granite but too fine-grained for the minerals to be seen, even with a hand lens. Small phenocrysts of quartz and feldspar may be present. This specimen from Glencoe, Scotland, shows alternating bands which are characteristic of this lava. (K)

Porphyritic pitchstone is dark glassy volcanic rock with many small phenocrysts of feldspar and quartz. It is found in dykes or sills near granitic centres or in lava flows. The phenocrysts probably formed before the lava was erupted and became incorporated in the rock when it was suddenly frozen after eruption. (K)

Syenites and Diorites

Larvikite A polished slab, such as that used for ornamental purposes, is illustrated. The beauty of the rock is due to the blue schillerization displayed by the feldspar crystals. The rock has a composition similar to syenite. The ferro-magnesian minerals are not obvious in a hand specimen since they occur as small olivine, magnetite and titanaugite clots. (CP)

Diorite has a characteristically speckled appearance. It is coarse- or medium-grained and of intermediate composition, having less than 10 per cent free quartz and oligoclase to andesine plagioclase. The ferro-magnesians are typically amphiboles but biotite and pyroxene can be present. Sphene is a common accessory, as are apatite and magnetite. The texture, as in the specimens shown, is usually equigranular. When diorites become richer in quartz and

Larvikite

Diorite

Syenite

Porphyritic micro-syenite

Porphyritic andesite

Nepheline syenite

Diorite

alkali feldspars they grade into granodiorites; when labradorite becomes the dominant plagioclase diorite grades into gabbro. Diorite is usually found in plugs and as marginal masses to plutons. (C, CP)

Syenite is an intermediate coarse-grained rock, often with a pinkish or grey overall colouring. There is less than 10 per cent free quartz (with more, it grades into granite). The main minerals are alkali feldspar, amphibole or pyroxene, and biotite. Syenite is an awkward rock to define; a number of varieties have been named, for example **quartz-syenite** which is over saturated with respect to total silica, and **nepheline syenite** which contains the feldspathoid nepheline. Syenites are equigranular and only rarely porphyritic. They occur in plutonic masses as well as in sills and dykes. (K)

Porphyritic micro-syenite is a medium-grained equivalent of syenite. The pale phenocrysts in this specimen are of feldspar. This group of rocks also includes the rhomb-porphyrys of Norway which are so characteristic of that area that they are useful as indicators of glacial movement (marker erratics). (CP)

Porphyritic andesite has a fine-grained matrix with a composition similar to diorite. When examined microscopically (the groundmass is too fine-grained to be seen well with a hand lens), plagioclase ranging from oligoclase to andesine can be identified with amphibole and biotite. In this specimen phenocrysts of pale feldspar and darker amphibole are visible. Andesites are generally not as dark as basalts and are erupted from rather violent volcanoes. (K)

White trachyte

Vesicular andesite

Andesitic pumice

Andesite

Trachytic agglomerate

Andesite from Mauna Kea, Hawaii

Trachyte

Dark andesite

Pink trachyte

Porphyritic andesite

Non-porphyritic trachyte

Poas, Costa Rica

Syenite seen at a magnification of ×20

Syenite seen at a magnification of x 20. This specimen from Bamle, Norway, can be designated hornblende syenite and in the lower part of the slide a large mass of hornblende, showing as browns and purple-blue shades, is visible. More hornblende is to the top left of the specimen. The black opaque patches are iron oxide. In the upper part of the slide there are pale grey prismatic crystals of plagioclase which have thin dark and light stripes because of repeated twinning.

Intermediate Lavas

Trachyte is composed predominantly of up to 10 per cent quartz, plagioclase (albite, sanidine or sodium-potassium varieties) and ferro-magnesians such as biotite, hornblende and

augite. The texture is fine-grained but phenocrysts of plagioclase, often sanidine, commonly occur to give a porphyritic texture. Trachytic texture is typical and consists of an alignment of the feldspar laths parallel to each other due to the flow of the molten lava. The overall colour of trachytes varies from white or grey to pink and brown. With an increase in quartz content they grade into rhyolites; when calcium-rich plagioclase increases they become alkali basalts.

Andesite has a similar composition to diorite. It therefore contains plagioclase in the range oligoclase to andesine, with pyroxene, amphibole and biotite and less than 10 per cent quartz. Andesites are often porphyritic with phenocrysts of feldspar or pyroxene and can be glassy, vesicular or amygdaloidal. They are usually darker than rhyolites being brown, purple, green or greyish in colour.

Andesitic pumice is a highly vesicular or scoriaceous rock so full of gas-bubble cavities that it may float on water. This specimen is from a volcanic centre of Pleistocene age at Puy de Parion in the Auvergne region of France. (L)

White trachyte has much pale feldspar and nearly 10 per cent pale quartz. This specimen from the Carboniferous strata of central Scotland exhibits a porphyritic texture. (L)

Vesicular andesite   Here a specimen from a recent eruption of the volcano Poas in Central America has numerous gas-bubble cavities which as yet are unfilled. (CP)

Andesite   A fine-grained, pinkish, porphyritic specimen with phenocrysts of pale plagioclase. Some of the pink colouring is from hematite-rich sediments which overlie the Ordovician lava flow from which the specimen comes, in West Cumbria, UK. (CP)

Trachytic agglomerate contains fragments of country rock and trachyte lava set in a matrix of lava. Such rocks are found in and around volcanic craters. The specimen is from Italy. (L)

Andesite from Mauna Kea, Hawaii   This specimen from a recent eruption shows porphyritic texture with small phenocrysts of pyroxene. (CP)

Trachyte   This specimen from East Germany has a porphyritic texture with phenocrysts of pale feldspar set in a fine matrix. (K)

Poas is an active andesite volcano in Costa Rica. The large crater is occupied by an acidic lake and in the cliff beyond the water stratified ash and lava deposits can be seen. Active fumaroles issue gas to the right of the lake. Andesite volcanoes erupt with violence and much of the material from this crater is highly vesicular.

Dark andesite   The surface of a rock specimen is often best seen when it has been cut and polished, as in this specimen from North Wales. Even when the specimen being examined is uncut, a film of water will give a similar effect. Phenocrysts of pyroxene and feldspar give this example its porphyritic texture. (K)

Pink trachyte has this overall colour because of the colour of the feldspar present in the matrix. Some dark phenocrysts of amphibole are visible. The specimen is from Peru. (L)

Porphyritic andesite   This specimen from Poas in Central America has some quite large phenocrysts of pyroxene. It also tends to be vesicular. (CP)

Non-porphyritic trachyte   A specimen from near the Tay Bridge in central Scotland. (K)

Olivine gabbro

Leucogabbro

Gabbro from the Skærgaard
complex, Greenland

Gabbro

Serpentinite

Dunite

Serpentinite

Bojite

Peridotite

The Cuillin Hills, Isle of Skye

Basic and Ultra-basic Rocks

Gabbro is a coarse-grained, dark-coloured rock of basic composition which occurs in large intrusions. These tend to be complex sheet- or saucer-shaped masses, often with a number of layers. Other basic and ultra-basic rocks also occur in such bodies, as well as in sills and dykes. Gabbro frequently occurs in layered complexes, as at Skaergaard in Greenland. Rich in calcium plagioclase (labradorite to anorthite), gabbro also contains pyroxene and olivine. Quartz, hornblende, biotite and magnetite can occur in small amounts too. With increasing ferro-magnesian minerals gabbro grades into peridotite, and if the feldspar content increases significantly it becomes anorthosite. Gabbros tend to be equigranular, with phenocrysts being unusual. Ophitic texture is, however, common, with euhedral lath-shaped feldspars enclosed in pyroxene. Dolerite and basalt are chemically similar to gabbro, merely being of medium and fine grain size respectively.

Peridotite is the most common rock of ultra-basic composition. It is generally coarse-grained, of granular texture and dark coloured. It may be a virtually mono-mineralic rock with the dominant constituent being olivine. Pyroxene, amphibole and garnet may be present. When feldspar occurs peridotite grades into picrite, which consists of ferro-magnesian minerals plus about 10 per cent feldspar. Layered structures are common in peridotites, which often occur in association with gabbros, and possibly form by differentiation from basic magmas. A wide variety of ultra-basic rocks has

been named, usually by reference to the dominant mineral, for example olivinite, pyroxenite, serpentinite and hornblendite. The total silica content of these rocks is less than 45 per cent and minerals which contain chromium, platinum and related elements are virtually confined to ultra-basic rocks.

Olivine gabbro is a dark, coarse rock with a blotching of green and white. The pale grains are plagioclase, the dark ones augite. Pale green olivine is seen as granular particles, especially when a hand lens is used. The specimen is from the Isle of Skye, Scotland. (K)

Leucogabbro is a paler than normal gabbro, which here contains more feldspar than average, with hornblende tending to become dominant over augite. There is also interstitial quartz and orthoclase. The specimen is from the Carrock Fell complex in Cumbria, UK. (CP)

Gabbro from the Skaergaard complex, Greenland. This specimen contains large, dark pyroxene crystals randomly orientated among pale plagioclase. Small granular crystals of olivine are also present. The overall structure of the intrusion from which this specimen comes resembles a pile of (concave-upwards) saucers. (CP)

Gabbro A specimen from the Insch intrusion in the Grampians of Scotland shows the typical mottled texture with pale plagioclase and dark pyroxene. (L)

Serpentinite is a dark rock often grey, reddish or green in colour in patches and veins. It can be

coarse- to medium-grained and contains serpentine minerals such as fibrous chrysotile and platy antigorite, with olivine, magnetite and other ferro-magnesians. It may form from original magma or by the alteration, involving hydration, of ultra-basic rocks. (CP)

Dunite is a granular ultra-basic rock which can vary in colour from green to brownish. This specimen shows the characteristic saccharoidal appearance. Because dunite is virtually mono-minerallic, containing over 90 per cent olivine, the name olivinite is often used. Chromite is an accessory mineral and dunite is frequently associated with important deposits of chrome and platinum. The specimen is from North Carolina, USA. (K)

Bojite is a strikingly coarse-grained gabbroic rock with large patches of plagioclase and primary dark green amphibole and some pyroxene. Magnetite is a common accessory mineral. The specimen is from Angmagssalik, Greenland. (K)

Peridotite is a dark heavy rock with a granular texture. It is composed of much olivine with other ferro-magnesian minerals. The specimen is from north-west Scotland. (K)

The Cuillin Hills on the Isle of Skye, off the west coast of Scotland, are largely composed of gabbro, which has weathered into sharp, jagged peaks, beloved of mountaineers. This view from Elgol, looking north-west to the Cuillins, also shows the weathered White Sandstone of middle Jurassic age.

IGNEOUS ROCKS

Anorthosite

Olivine basalt

Dolerite

Troctolite

Basalt

Amygdaloidal basalt

Weathered dolerite

Amygdaloidal olivine basalt

Dolerite

Porphyritic basalt

Norite

Spilite

Anorthosite is a coarse-grained basic, plutonic rock made of more than 80 per cent plagioclase, usually labradorite. Ferro-magnesian minerals make up the remainder of the rock. This specimen has areas covered with pale plagioclase, and the ferro-magnesian parts are of interest because the dark pyroxene has reaction rims of reddish garnet. Anorthosite occurs in plug-shaped intrusions, as sheets and patches in basic intrusions. The rock has been found on the Moon, where it is dated at over 4,000 million years. (K)

Medium/Fine-grained Basic Rocks

Olivine basalt is a dark, fine-grained volcanic rock with many vesicles up to 2 mm in diameter. In this specimen from Gran Canaria, Canary Islands, there are large patches of pale green olivine. The texture is too fine-grained for the minerals in the matrix to be seen with a hand lens. (C)

Dolerite has the same chemistry as gabbro and basalt but because it is medium-grained the constituent minerals, pale plagioclase and darker pyroxene, give the rock a speckled appearance. Other minerals are quartz (less than 10 per cent), olivine and magnetite. Dolerite cools more slowly than basalt and forms commonly in hypabyssal intrusions such as sills and dykes. (CP)

Troctolite is a dark mottled rock which has been called 'troutstone' because of its colouring. Chemically it is related to gabbro and dolerite, but lacks pyroxene and is rich in grey plagioclase and brownish olivine. When examined microscopically the olivine is seen to have rims of pyroxene. (K)

Basalt is a dull fine-grained rock with an even texture. This specimen lacks features such as vesicles and phenocrysts. When examined microscopically the small laths of plagioclase and crystals of pyroxene are visible with some olivine and magnetite. The specimen is from Mount Etna, Sicily. (CP)

Weathered dolerite This specimen shows the very common spheroidal weathering typical of fine- and medium-grained basic igneous rocks. The rock here is a porphyritic dolerite from Skye, Scotland. It has a 'rusty' oxidized surface and skins have become separated around a central core where water has seeped into joint systems. Slightly acid rain water attacks the minerals in the dolerite and reduces a thin layer to bicarbonates and clay which are readily dissolved. (CP)

Amygdaloidal basalt This specimen contains many rounded patches, up to 5 mm in diameter. These are amygdales, mainly of calcite, which occupy vesicular cavities. The specimen is from Antrim, Northern Ireland. (L)

Dolerite showing two distinct grain sizes, the central band being of gabbroic texture. In this portion the plagioclase and augite are readily identified. (CP)

Amygdaloidal olivine basalt A specimen with a greenish fine-grained matrix and a rusty weathered upper surface. The amygdales are

The Whin Sill

Basalt in thin section

Dolerite in thin section

both round and elongated and contain calcite, the rhombic cleavage of which is visible. The matrix consists of plagioclase, pyroxene and olivine, with some quartz. Vesicular and amygdaloidal basalts are common in the upper parts of lava flows. The specimen is from Cumbria, UK. (CP)

Porphyritic basalt A dark fine-grained matrix with striking greyish-green phenocrysts, up to 1 cm long, of plagioclase feldspar. The euhedral, tabular phenocrysts must have formed when the basaltic magma was liquid, before eruption. (CP)

Norite has a medium- to coarse-grained texture and gabbroic composition. However, olivine is virtually absent and orthopyroxene is abundant. In gabbro and dolerite clinopyroxene (augite) is dominant. (K)

Spilite is a fine-grained basic rock which often, as in this specimen, has a vesicular or

amygdaloidal texture. The pale amygdales here are of calcite, though quartz is often present. Spilites contain abite (sodium-rich plagioclase) and augite, or more often its altered equivalents such as chlorite. Spilite occurs in pillow lavas interbedded with deep marine sediments and though there is debate as to their exact origins it is often asserted that their chemistry is due to reaction between basaltic lava and sea water. The specimen is from Nassau, the Bahamas. (C)

The Whin Sill is an extensive sheet of dolerite which occurs within the sedimentary strata of Carboniferous age in north-east England. The rock is dark and medium-grained and has been radiometrically dated at 295 million years. As is the case with many sills, the Whin Sill transgresses from one horizon to another, notably in the Cross Fell area. Its resistance to weathering and erosion produces steep sea and inland cliffs. In this exposure on the coast of Northumberland the vertical columnar jointing has been whitened by nesting seabirds.

Basalt in thin section, here magnified × 30, is seen as a mass of crystals which are not as well formed as those in the section of dolerite. This is probably due to the more rapid cooling of the basalt, which has caused the crystals to be fine-grained. The grey 'lath-shaped' crystals are plagioclase feldspar, while the browns, blues and greens are pyroxene. On the left of the slide there are a few rounded, brightly coloured grains of olivine, and black opaque iron oxide grains are also present.

Dolerite in thin section (× 30) is one of the most attractive of rocks, and is amazingly different from the dull speckled rock that appears in the field. When compared with the section of basalt the larger, better formed crystals can be seen. This specimen shows much prismatic plagioclase, with the twinning represented by alternate dark and light grey colours in the crystals. The brightly coloured grains are pyroxene.

Pyroclastics

Volcanic bomb This specimen from Kilauea, Hawaii, shows the typical rounded and slightly spindle-shaped structure which develops as the ballistic fragment of lava solidifies during flight. (K)

Pumice is usually produced by explosive activity from volcanoes rich in rhyolitic or andesitic lava. It is a highly vesicular material made in great quantities when foaming gas-rich lava freezes. The specimen is from New Zealand. (K)

Ropy lava, pahoehoe forms when basaltic lava continues to flow beneath a solidified but plastic crust. The flowing lava stretches the crust, producing ropes and folds on the upper surface. Gas release is gentle during such movement and the lava is vesicular. The term 'pahoehoe' is the Hawaiian word for the rock. The small specimen in the centre of the page is from Iceland, the one at the bottom left is from Gran Canaria, Canary Islands, and the other two are from Mauna Loa, Hawaii. (CP, C).

Spatter As lava cools many shapes are produced. When miniature eruptions occur through a mass of mobile gas-rich lava, spatter cones develop ejecting 'spatter' as frothy clots of lava. (K)

Agglomerate is a pyroclastic rock of coarse grain size, containing fragments which are generally over 2 cm in diameter. The fragments are of lava, which is often highly vesicular, and country rock from around and below the crater which has become caught up in the eruption. The particles are mainly angular but some spindle-shaped clots of lava may be found. Agglomerates occur near and in volcanic craters. (C, L)

Tuff is consolidated volcanic ash and can occur in neat strata, especially when deposited under water. Graded bedding and other sedimentary structures can also occur. The particles are fine-grained, and are usually far less than 2 mm in diameter. Ash can be made of a great variety of materials, including small lava fragments, crystals of feldspars and ferro-magnesian minerals, glass and rock fragments. During very explosive activity the finest ash and dust can be taken high into the atmosphere. The ash from Mt St Helens, which erupted in 1980, reached 20 km into the atmosphere, and the eruption of Tambora in Indonesia in 1815 carried ash and dust as high as 50 km. Materials ejected like this are carried by prevailing winds, and as the coarser grains fall first a tapering 'dune' of tuff is formed with the volcano at the deeper end. Wind systems can carry volcanic dust thousands of kilometres, high in the atmosphere, and the geological record contains many thin 'marker-horizons' of ash and clay formed from ash falls. (C, L)

Pelée's hair consists of fine threads of basalt which have been formed when lava spray is drawn out by the wind as it issues through small orifices. The mythical Pelée, according to Hawaiian legend, is a lady who lives inside the volcano Kilauea. (CP)

Volcanic tuff In thin section at a magnification of × 20 volcanic tuff appears as a mass of irregular fragments welded into a drawn-out mass of lava, in this example andesite.

Agglomerate

Agglomerate

Agglomerate

Tuff

Tuff

Volcanic bomb

Ropy lava

Ropy lava

Pumice

Ropy lava

Ropy lava

Spatter

Pelée's hair

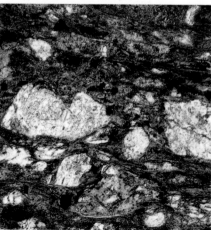

Volcanic tuff

Metamorphic Rocks

Metamorphism involves the alteration of pre-formed rocks by pressure, temperature and migrating fluids, often in environments deep in the Earth's crust. New characteristics, reflecting these conditions, are 'printed' over the existing features of the rock. Because of the severe conditions which rocks undergo during metamorphism, the minerals of which they are originally made may become unstable and change to maintain equilibrium with the new environment. Metamorphism does not involve melting, though at very high pressure and temperature rocks may become plastic. The temperature involved may be as high as 700°C, and hydrostatic or confining pressure, which increases with depth, may be as high as 6,500 atmospheres over 20 km below the surface. Stress, directed pressure produced by folding and other movements in the Earth's crust, creates new fabrics such as lineation and cleavage which replace original structure like sedimentary bedding. Fluids circulating through the rocks at depth allow chemicals to be introduced and taken away from rock systems. Three main types of metamorphism are recognized. **Regional metamorphism** results from large-scale folding and burial of pre-formed rock, usually in areas of orogeny. **Contact metamorphism** is produced by direct heating of rocks which are next to intrusive magma, or immediately beneath a lava flow. **Dynamic metamorphism** occurs where large-scale faulting breaks and deforms rocks next to the fault plane. Temperatures and pressures are greater here than along smaller faults higher in the crust, where rocks are more brittle.

Slate   At the lowest grade of regional metamorphism, where temperatures and pressures are low, only rocks like shale, mudstone and volcanic ash are altered. The metamorphic rock produced is slate, a fine-grained rock with cleavage which allows it to be easily split into thin sheets. This specimen has been broken along a cleavage plane and running diagonally across this surface are traces of the original bedding.

Gneiss   At the extreme high-grade end of regional metamorphic processes gneiss is formed. This example shows the coarse-grained, granular nature of the rock. The pale minerals are quartz and feldspar, which alternate with darker bands of amphibole and biotite mica.

Dolerite intrusion in shales and sandstone
The effects of contact metamorphism depend on the size of the igneous body, the type of country rock and the type of magma, or lava, producing the heat. Here a small intrusion of dolerite, showing typical blocky structure and rusty weathering, has intruded shales and sandstones. Right next to the dolerite a layer of originally dark shale has been 'baked' and whitened. The sandstone at the top of the cliff face has not been affected by heat from this intrusion and retains its original features.

Dolerite intrusion in shales and sandstone

Slate

Gneiss

Slate

Phyllite

Phyllite

Slate with bedding

Slate

Slate

Slate with pyrite

Regional Metamorphism

Regional metamorphism includes a number of processes which bring about profound changes in pre-existing rocks, over very wide areas of the Earth's crust. Vast tracts of Pre-Cambrian basement rocks, along with those involved in more recent orogenic activity, have undergone regional metamorphism. The mobile belts of the crust, where mountain building takes place, are the regions where temperatures and pressures are sufficient to alter pre-formed rocks. The depth at which the changes are brought about greatly influences the degree of metamorphism. At great depths migrating fluids are able to facilitate metamorphic changes by adding chemicals to the rock systems. At the highest grades of alteration rocks may become plastic, and here the border between metamorphic and igneous rocks is approached. Regionally metamorphosed rocks tend to show a strong preferred orientation of minerals which gives the new rock textural features such as cleavage, schistosity and gneissose banding. Characteristic mineral assemblages develop under particular conditions of temperature and pressure, and an understanding of these helps geologists to unravel the processes which have been involved.

Low-grade Regional Metamorphism

Slate is a rock which is easily split into thin sheets because of well-developed cleavage produced by the parallel alignment of minute flaky minerals, like mica and chlorite, in response to pressure conditions acting on the rock. Slate is produced from fine-grained pelitic sediments such as clay, mudstone and shale, as well as fine-grained volcanic tuff. Bedding and other sedimentary structures are often still visible, and fossils, though deformed, also occur. The cleavage can be at any angle to the original bedding and usually develops at right angles to the pressure directions. When alternating beds of, for example, shale and sandstone are metamorphosed at low grade, the shale beds develop cleavage while the more resistant sandstones do not. The minerals making up slate are fine-grained and so are not visible to the naked eye. They are micas, quartz, clay minerals and other constituents related to the original rock. Porphyroblasts of euhedral iron pyrites are common. Slates vary in colour from black through grey to green. The specimens illustrate the colour variety; one shows original bedding (*left centre*), one shows organic material (*bottom left*) and one shows pyrite porphyroblasts (*bottom right*). (CP)

Phyllite This rock falls between slate and schist in metamorphic grade. Phyllites are slightly coarser grained than slates and wavy foliation replaces cleavage. A separation of minerals can take place which produces bands rich in mica and others rich in quartz and feldspar. These minerals are, however, of small grain size and a hand lens is needed to detect details. The dominance of mica in this rock gives it a characteristic sheen. Fine-grained pelitic sediments and slates will be altered by the conditions which form phyllite. These rocks range in colour from dark grey to green, particularly when chlorite is present. (K)

Slate showing cleavage planes

Slate showing cleavage planes Slates are characterized in the field by cleavage planes, produced by the pressure conditions of metamorphism. These can intersect at any angle with the original bedding planes. Here steeply dipping slates show both cleavage and bedding. The cleavage surfaces appear as sharp edges projecting from the cliff, whilst an original bedding plane lies very steeply as a smooth surface at the right-hand edge of the picture.

Slate in thin section At high magnification, slate appears as a very fine mass of quartz and mica. The quartz grains are grey or black while the mica flakes are brightly coloured. This rock is very similar to the shale or mudstone which it was before metamorphism, but there is some preferred orientation of the grains due to metamorphism.

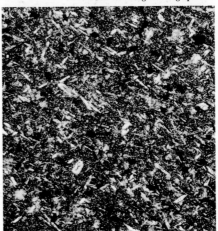

Slate in thin section

27

Biotite schist

Hornblende schist

Garnet mica schist

Marble

Muscovite schist

Kyanite schist

Kyanite schist

Muscovite chlorite schist

Fold mountains

Medium-grade Regional Metamorphism

A variety of metamorphic conditions is present at this grade; temperatures and pressures are higher than for slates, and the detailed mineralogy of schists and other rocks which develop at medium grade indicates the precise chemical and physical conditions in which metamorphism took place. In the Scottish Highlands zones of progressive regional metamorphism can be mapped by using index minerals in the schists. Chlorite schists (the least altered) grade into rocks rich in biotite, kyanite and sillimanite with increasing metamorphism. Schist is the typical rock of this medium-grade metamorphism. It is a rock characterized by a wavy foliation, or schistosity, caused by the alignment of mica crystals. Other minerals include quartz and feldspar. Ferro-magnesian minerals can be present depending on the parent rock's composition. Porphyroblasts of garnet are common in schists but these are not often of any great size. Schist grade rocks which have developed from basic igneous material tend to have a granular texture and are termed amphibolites. When limestones suffer medium-grade metamorphism, marbles are produced which are noted for their parallel alignment of minerals and the presence of calc-silicate minerals developed from impurities in the original limestone. Quartz and clay in limestone may produce diopside, garnet and olivine during metamorphism. Schists and related rocks are formed near the centres of orogenic zones at considerable depth in the crust. Many rocks can be altered in these regions, including pelitic sediments, slates, phyllites, arenaceous sediments and some igneous rocks.

Biotite schist is a dark medium-grained rock with abundant black biotite mica on the foliation surfaces. Quartz and feldspar are also present. This type of rock develops from pelitic sediments. (K)

Hornblende schist is characteristically greenish in colour and has some feldspar and garnet among the silvery-green amphibole. Foliation is well developed; the original rock was probably a basic igneous material such as dolerite. (CP)

Garnet mica schist is a medium-grained muscovite-rich rock with porphyroblasts of dark garnet visible on the foliation surfaces. Other minerals in this specimen include quartz and chlorite. The presence of garnet indicates a higher level of metamorphism than that undergone by the biotite schist. (K)

Muscovite schist is a much paler rock than biotite schist because of the abundance of silvery-white muscovite mica. This example also contains quartz and feldspar. (CP)

Marble Because of the alignment of the darker calc-silicate minerals in this example it is possible to suggest that the rock has formed as a result of regional metamorphism. The rock is predominantly made of calcite and therefore can be easily scratched by a knife blade. It also effervesces when cold dilute hydrochloric acid is poured over it. (K)

Kyanite schist The presence of kyanite indicates a high level of metamorphism and these rocks can be coarse- or medium-grained. Two specimens are shown; the darker one has much biotite mica, the paler grey example has muscovite mica. Both also contain quartz and feldspar, and the paler specimen has some

Mica schist in thin section

porphyroblasts of garnet as well as small-bladed crystals of kyanite which lie parallel to the schistosity. (K)

Muscovite chlorite schist This specimen shows wavy foliation surfaces which are rich in mica and chlorite, the latter giving the greenish sheen. Biotite and quartz are also present. (K)

Fold mountains, such as these ice-covered peaks on the French-Italian border, have in their roots the temperature and pressure conditions which produce metamorphism of schist grade. Only when much weathering and erosion has taken place do rocks like schist appear on the Earth's surface.

Mica schist This thin section shows a coarser grain size than slate, and a more obvious foliation. The pale and grey granular particles are quartz, while the bright greenish-brown and darker shades are crystals of mica which follow the foliation diagonally across the slide.

Granulite

Eclogite from Greenland

Eclogite

Augen gneiss

Gneiss

Gneiss from the Lewisian complex

A granular textured gneiss

Amphibolite

Eclogite

Amphibolite from Sutherland, Scotland

High-grade Regional Metamorphism

Rocks like gneiss and eclogite are formed under conditions of extremely high temperature and pressure, in the deepest regions of orogenic belts. This means that they are only exposed on the Earth's surface after much weathering and erosion has removed the overlying material. In these deep regions of the crust chemical mobility and the presence of pore fluids produce extreme changes in all rocks. The segregation of certain groups of minerals takes place and this gives gneiss its characteristic banding, where quartz/feldspar bands alternate with bands rich in ferro-magnesian minerals like biotite and hornblende. Gneisses are coarse-grained granular textured rocks which can develop from a wide variety of both sedimentary and igneous material. The banding in this rock can vary from a millimetre or so to a metre in width, and this structure is far less regular than cleavage or schistosity in less altered rocks, often being highly contorted in a way suggestive of a plastic phase during the rock's development. Gneisses make up much of the Pre-Cambrian basement formations in areas like the Canadian Shield, Greenland, north-west Scotland and Russia. Gneisses often have a granitic composition (though they can vary from acid through to basic), but eclogites, which form under similar high-grade conditions, have a bulk chemistry of a basic igneous rock, containing much pyroxene (often omphacite) and garnet in a granular coarse-grained texture. The high specific gravity of eclogite may suggest that it originated at very great depth, a fact that is further indicated by its occurrence in structural situations like diamond pipes and major fault zones.

Gneiss in thin section is very much coarser grained than schist, with only a few crystals filling the field of view. The crystals here show preferred orientation. In many gneisses the darker bands are rich in amphibole, and this example contains large crystals of hornblende which appear as brightly coloured masses, some showing typical cleavage traces. Some grey quartz grains are also present.

Eclogite from Greenland This specimen shows the characteristic 'mottled' appearance of red garnet and green pyroxene. (K)

Granulite is a dark coarse-grained rock with only a moderate degree of preferred orientation of minerals. This specimen from northern Scotland contains pyroxene, feldspar and some quartz. Granulites are formed under similar conditions of temperature and pressure to eclogite. (K)

Eclogite This specimen contains garnet, pyroxene and amphibole. It is from Ayrshire in southern Scotland. (K)

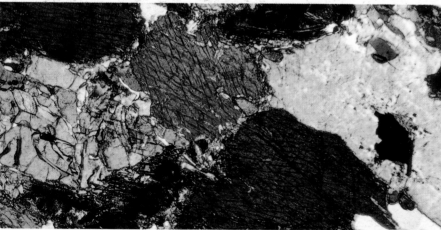

Gneiss in thin section

Gneiss landscape

Augen gneiss Some gneisses develop clots or 'eyes' of minerals such as feldspar, as in this specimen from Austria. The gneissose banding is thus modified and no longer distinct. Greenish amphibole, red garnet and pale quartz and feldspar are present in this example. (CP)

Gneiss This specimen from northern Canada shows the characteristic dark bands of biotite and amphibole alternating with pale bands rich in quartz and pink orthoclase. (CP)

Gneiss from the Lewisian complex of north-west Scotland has neat bands of dark biotite alternating with quartz- and feldspar-rich areas. (CP)

A granular textured gneiss in which the banding is not as pronounced as in the other specimens. There are sub-parallel bands of dark biotite mica set among much grey anhedral quartz and some feldspar. (CP)

Amphibolite This specimen is a coarse-grained rock with reasonably well developed foliation which is picked out by the mass of amphibole crystals. Much of this amphibole is actinolite; garnet and some feldspar are also present. Such rocks can develop by the high-grade metamorphism of basic igneous material. (K)

Eclogite A very dark, dense specimen composed almost exclusively of pyroxene with red garnet porphyroblasts. The specimen is from northern Norway. (K)

Amphibolite from Sutherland, Scotland A specimen full of black prismatic hornblende, which is orientated with the foliation of the rock. It is coarse-grained and has some feldspar and pyroxene in small amounts. (CP)

Folded gneiss

Gneiss landscape Much of the Earth's ancient continental crust is made of high-grade regionally metamorphosed gneiss, as here in the north-west of Scotland. During the Pre-Cambrian era, when these rocks were formed, heat flow from the interior of the Earth was probably much greater than today and so very extensive regions of gneiss were made. Though gneisses are probably formed at great depth below mountain chains today, the extensive conditions of the Pre-Cambrian era have probably never been repeated. The coarse banded gneisses produce a rough hummocky landscape. Similar terrains occur in Greenland, Canada and the USSR.

Folded gneiss In extreme conditions of regional metamorphism rocks may become plastic, and tight folding, as in this example, can take place. Here gneiss with characteristic alternating dark and light bands is exposed on a wave-cut platform. The distance across the picture is only 2 m. Rocks such as this where there is a close association of acidic (pale, quartz- and feldspar-rich bands) and basic (dark, ferro-magnesian-rich layers) are called migmatites.

Spotted rock

Marble

Marble

Hornfels

Chiastolite slate

Metaquartzite

'Sugary' marble

Hornfels

Grey marble

Dynamic metamorphism

Contact or Thermal Metamorphism/Dynamic Metamorphism

When igneous rock in the form of lava or magma comes into contact with pre-formed country rock its heat permeates and changes this rock. The actual temperatures against the intrusion or lava flow may approach 1000°C. The temperature gradient away from the heat source may decline very steeply from the margin of a small body such as a dyke, with limited metamorphism, but a large granite pluton will produce changes in the country rocks for many kilometres around it. The extent of the metamorphic aureole (the region in which metamorphism occurs) depends on the nature of the country rocks as well as the size and nature of the igneous material. Sandstones and quartz-rich sediments, for example, are less profoundly altered than fine-grained pelitic sediments which contain much clay. The type of igneous rock causing the metamorphism also influences how much and in what ways the country rock is changed. Basic igneous rocks tend to be hotter than acid rocks, while the latter contain more fluids which can penetrate surrounding materials carrying new elements into the country rocks. In reasonably consistent country rocks affected by heat from a large intrusion, zones of increasing metamorphism can be mapped by using characteristic metamorphic minerals, such as chiastolite, andalusite and cordierite, which develop in response to the increase in temperature towards the intrusion.

Spotted rock This specimen is of a fine-grained pelitic sediment from the outer parts of a granite aureole. Some of the characteristics of the original rock remain and dark spots of cordierite have developed. (CP)

Marble This crystalline calcareous rock has developed by the intrusion of syenite into dolomitic limestone. Magnesium in the dolomite has aided the development of brucite, seen as bluish-green patches. (K)

Marble from the aureole of a granitic intrusion. This marble is a crystalline calcite rock containing both serpentine and forsterite as greenish blotches which have developed from impurities in the original limestone. (K)

Chiastolite slate This rock has suffered both regional and thermal metamorphism. Initially fine-grained pelitic sediment, it was converted to slate by regional metamorphism and subsequently intruded by granite. The cleavage of the slate is still present but prismatic crystals of chiastolite have developed due to heat from the granite. (CP)

Hornfels In close proximity to a granite intrusion rocks become profoundly altered. This rock was originally volcanic tuff but is now a hard, flinty hornfels rich in brownish-red garnet, quartz and feldspar. (CP)

Metaquartzite Quartz-rich sediments become recrystallized, and pore spaces are filled with new quartz growing around the original grains.

Marble in thin section

Mylonite in thin section

A completely interlocking fused mosaic of quartz crystals is thus produced. The new rock may lack original bedding and sedimentary structures depending on the degree of recrystallization. (CP)

'Sugary' marble A very coarse-grained rock with a granular texture, developed by the heat from a dolerite sill altering reasonably pure limestone. (CP)

Hornfels A fine-grained crystalline rock with none of the original sedimentary features remaining. It is a quartz-rich rock with biotite mica having developed during metamorphism. This specimen is from a granite aureole. (K)

Grey marble This marble has developed from a dolomitic limestone and is rich in greenish olivine and dark serpentine which appear as streaks through the rock. These minerals have developed from magnesium and other impurities in the original limestone. (K)

Marble in thin section from Carrara in Italy shows an interlocking mass of calcite developed from limestone by the heat of contact metamorphism. The rock is composed completely of calcite, which shows characteristic cleavage traces and bright pale colours produced by polarized light passing through the thin slice of rock.

Dynamic metamorphism is associated with large-scale fault movement, particularly thrusting. Rocks along the thrust plane are pulverized and the minerals may be streaked out by the lateral movement of the fault. In this photograph a major thrust plane is well exposed in the hillside beyond the sea-loch. Here gneisses, forming the upper part of the hill, have been thrust over younger strata, with a very sharp contact between them. Next to such a fault plane a rock called mylonite, made of the pulverized and streaked-out fragments, is found.

Mylonite When magnified (here at about × 15 life size) the elongated fragments in this mylonite reflect the dynamic conditions of metamorphism which occur along a major thrust plane. (RMS)

Sedimentary Rocks

Sedimentary rocks are those formed on the Earth's surface in environments familiar to us, so their formation is easier to assess than that of igneous or metamorphic rocks formed at great depth in the crust. By using the principle of uniformitarianism, which suggests that we can understand and interpret rocks formed in the past by reference to processes happening today, it is possible to reconstruct ancient environments of sedimentation. Sedimentary rocks are characterized by bedding planes or stratification. This feature is a series of roughly parallel plane surfaces representing the original surfaces on which the sediment was deposited. Most of the sediments which are preserved as rocks were formed in marine environments, though much deltaic and desert sediment also occurs in the geological record.

Sedimentary rocks are secondary, in that they contain grains derived from earlier formed rocks. This recycling process occurs constantly, and igneous, metamorphic and earlier sedimentary rocks can all suffer the processes of weathering, erosion, transportation and deposition. This can take place in many systems including river, glacial, and aeolian environments. The total make-up of the rock, which includes the chemistry and shape of the grains, the bedding structures, fossil content and so on, is its *facies*. This sum total of the features of the rock contains the clues as to its formation. Sediments are the rocks which contain fossils and so are of great interest to palaeontologists. Many important resources are trapped in sedimentary strata, including oil, coal and iron ores.

After the sediment has been deposited it can be subjected to a variety of processes which convert the loose particles into solid rock. Diagenesis is the term for these processes which take place at relatively low temperatures and pressures near the Earth's surface. With increasing depth metamorphism takes over. Compaction and cementation are two of the most important diagenetic events. Compaction removes pore fluids (marine sediments can take considerable quantities of sea water with them

The Red Mountains, near Oak Leaf Canyon, Arizona, USA

as they become buried deep in the crust) and the grains are arranged with closer packing. The weight of overlying strata has a considerable effect on compaction. Cementation involves the formation of secondary minerals around the grains, often quartz, calcite or iron oxides. In limestones recrystallization may occur very soon after deposition to produce a rock without spaces between the grains.

Certain elements of structural geology are best displayed by sedimentary strata. Though beds of sediment are deposited on a more or less horizontal surface, these bedding planes can be folded and tilted by earth movements. The angle of inclination of such a surface from the horizontal is called the **dip**. As well as having its angle of inclination, dip also has a compass

direction, and a bearing taken at right angles to the direction of dip is called the **strike**.
Unconformities are another typical feature which involves sedimentary strata. These are breaks in the geological record often resulting from a period of erosion followed by the deposition of a new series of sediments. The plane of the unconformity will have characteristic features indicating how it was produced; for example, marine erosion tends to make smooth surfaces while old land surfaces are often irregular.

The three main groups of sediments are illustrated in detail on the following pages. These are the **detrital** or fragmental sediments which include conglomerates, sandstones and shales; the **organic** sediments which include

some limestones, coals and bone beds; and the **chemically formed** sediments including some limestones and the evaporites.

The Red Mountains near Oak Leaf Canyon, Arizona, USA, are formed from Triassic sandstones deposited by rivers and streams flowing over a vast plain. The bright colours of the strata are due mainly to iron oxides coating the sand grains. Modern weathering picks out the weaknesses in the strata, enlarging vertical joints to produce the characteristic stacks and chimneys on the crags.

Detrital sediments

Wave Rock at Hyden in Western Australia

Dip and strike

Dip and strike are well displayed by these beds of limestone and shale on the coast of Somerset, UK. The softer shale beds have been eroded more readily than the limestones leaving the latter as prominent landward-dipping strata. The true dip is the greatest angle that can be measured down these surfaces, and the strike is the compass bearing at right angles to this. In this picture the strike follows the direction of the 'mini-escarpments' towards the cliffs in the far distance.

An unconformity may be an old erosion surface with more recent sediments deposited over it. Here in the far north of Scotland an unconformity exists between the basement gneisses which form the low hummocky ground and the sandstone mountains rising in the distance. Originally many thousands of metres of this Torridonian sandstone covered the whole region, but only a few remnant outliers are now left. The irregular gneiss surface was the land some 900 million years ago, with the rivers which deposited the Torridonian flowing over it.

Detrital sediments are made of particles worn from pre-formed rocks. These particles of rounded quartz are from a sandstone of Triassic age. Modern sands with particles of this size and shape are characteristic of arid, wind-blown sands found in desert regions. By using the principle of uniformitarianism it is reasonable to suggest that these grains were formed by similar processes over 200 million years ago. (CP)

Wave Rock at Hyden in Western Australia is a spectacular example of the power of erosion. The cliff of Pre-Cambrian granite has been sculptured by wind-blown sand. The dark bands on the cliff are stains from mineral solutions.

The Petrified Forest, Arizona, USA.   This mound of bentonite clay consists mainly of very small particles of clay minerals like montmorillonite, and when wet it swells in size. When examined microscopically such clays are seen to contain silicate minerals typical of intermediate lavas. These clays may therefore have formed from volcanic ashes which have been devitrified and stratified.

An unconformity

The Petrified Forest, Arizona, USA

Boulder clay (till)

Calcareous breccia

Conglomerate

Breccia

Iron-rich conglomerate

Polygenetic conglomerate

A remnant patch of conglomerate resting on a cliff of gneiss

Beach deposits

Detrital Sedimentary Rocks

These are sedimentary rocks formed as a result of the processes of weathering, erosion, transportation and deposition. Thus they are made of fragments (clasts) broken off from pre-existing rock masses. The detailed chemistry and grain size of the detrital sediments depends on the nature of the original source material, the environment of deposition and post-depositional processes which convert the sedimentary particles into solid rock. All these factors interplay in a variety of ways which lead to the great diversity of detrital sediments. Because such sediments are forming today, by processes we can see and understand, it is possible to interpret ancient sedimentary rocks and suggest accurately how they were formed. Detrital sediments accumulate in vast thicknesses, especially in some marine environments such as on continental shelves, at the mouths of large river systems, and in ocean trenches. All these sediments have various structures related to the processes which have formed them. Bedding or stratification is the most obvious of these and originally this layering is quite flat and neat. It may be picked out by changes in grain size and sediment chemistry. Sedimentary particles which are deposited in a moving current of water or wind exhibit a structure called cross (or current) bedding, and in certain environments where sediment of differing grain sizes is allowed to settle through the water column, graded bedding will occur with the coarser grains lower in each bed than the finer particles. Generally the grain size of detrital sediments is determined by distance from the source area, and the power of the transporting medium. Boulders can be carried great distances by ice, but not by rivers, except possibly during flash flooding. The coarser sediment is therefore found in shallower waters and in environments like beaches and screes, and fine-grained clays are found far out at sea. This is only a guide, however, and many complications occur: for example, turbidity currents can carry reasonably coarse sediments out into deep marine environments, far away from land. The chemical composition of detrital sediments is related to the rocks in the source

area and to the chemical environments where the rocks are deposited and hardened. In general terms these rocks have a simple chemistry containing much quartz, because this mineral is both chemically stable and mechanically resistant. Feldspars are chemically less stable and their presence in, for example, a sandstone may indicate rapid burial or low levels of chemical weathering in the source area. Mica is common in some sandstones but in those where wind action has been involved this flaky mineral is absent, as it is easily blown away. Minerals like ferro-magnesians are easily weathered, and like feldspars they rot into clays and become incorporated into the detrital rocks. In the coarser-grained detrital sediments rock fragments are identifiable and it may be possible to trace these to a source area as in the case of erratics in glacial clays. Other minerals may be introduced after deposition and these may act as a cementing agent. Calcite, iron oxides like hematite, and glauconite are examples. Diagenesis is the term for the variety of processes which act near the Earth's surface, where temperatures and pressures are not high, to make hard rock from the unconsolidated sediment. Compaction, the elimination of water from pore spaces, and cementation are all involved.

Boulder clay (till) is a clay matrix made of ground-up rock material containing unsorted angular and sub-rounded rock fragments set with no preferred orientation. The fragments include igneous and sedimentary material and are derived from a wide area, though often boulder clays have a concentration of local material. The rock fragments are erratics and a study of their origins will indicate the direction of ice movement. When boulder clay has become indurated by diagenetic processes the name **Tillite** is used. Boulder clay is a common superficial deposit in areas which have suffered recent glaciation, including much of northern Europe, Britain and North America; some tillites are, however, of extreme geological age, and such rocks are found dating back to the Pre-Cambrian era. Because of their climatic associations, these deposits help with the determination of the former positions of land masses and hence how the continents have drifted. (CP)

Calcareous breccia is a coarse-grained detrital sediment made predominantly of angular particles. This shape suggests that water transport and erosion have not been important in the formation of the rock. In this example the fragments are of limestone and they are held in a brown lime-rich cement. (CP)

Conglomerate is a similar rock to breccia except that the fragments are rounded indicating water erosion. This example consists of grey quartz pebbles set into a grit and sand matrix. The size of the largest fragments suggests that a strong water current was involved. Such rocks often form on beaches and in very shallow water environments. (C)

Breccia A specimen containing very large, grey, angular limestone fragments held in a clay matrix rich in iron oxide. (CP)

Iron-rich conglomerate This example contains rounded quartz fragments and rock particles cemented by brown limonite. Cements such as this are common in detrital sediments and may indicate arid conditions. (C)

Polygenetic conglomerate A specimen containing a great variety of different rock fragments, including dolerite, sandstone and limestone held in a hematite-rich matrix. (CP)

A remnant patch of conglomerate resting on a cliff of gneiss. Conglomerates frequently form as the first layer above an erosion surface, with a plane of unconformity separating the older eroded series from the basal conglomerate of the newer series. This ancient cliff is probably still much as it was when the conglomerate (originally much more extensive) was heaped against it.

Beach deposits Breccias and conglomerates can result from the preservation of ancient beach deposits. Here in South Wales an old raised beach remains above the level of the modern beach, resting unconformably on a platform of steeply dipping Carboniferous limestones. The raised beach contains many fragments of limestone, and other rocks, cemented with calcite.

Black shale

Orthoquartzite

Greensand

Red sandstone

Arkose

Grit

Micaceous sandstone

Clay

Greywacke

Mudstone

Ripple-marked sandstone

The bedding surfaces of detrital sediments

Black shale is a fine-grained, thinly bedded rock with much quartz, mica and clay. Carbon and finely disseminated pyrite give the rock its dark colour. The high proportion of iron pyrites, which often replaces fossils in dark shales, suggests that these rocks are deposited in deep, possibly stagnant waters where the chemical environment is one of reducing conditions. The fossils in this example are free-swimming ammonites which lived above the sea bed, and bivalves which may have lived attached to floating algae. (CP)

Orthoquartzite This sediment has over 95 per cent quartz content, the original quartz grains being cemented by authigenic quartz which has formed around them. This has produced a tough, hard rock which nevertheless shows bedding and other structures. The specimen shown contains worm burrows. (CP)

Greensand is a quartz-rich sandstone with a high percentage of the green mica-mineral glauconite, occurring as flaky grains with a rounded outline among the quartz. Glauconite is a mineral typically found in marine sands, modern glauconitic sands being found near the margins of continental shelves. (K)

Red sandstone A specimen made up of medium-sized, rounded, well-sorted quartz grains. The red colouring is because of a hematite coating on the grains. This iron oxide acts as only a poor cement; the grains can easily be rubbed off with the fingers. The presence of iron oxide together with the rounding of the sand grains suggests deposition in an arid, wind-influenced environment. (K)

Arkose This specimen is of medium to coarse grain size and composed of quartz grains with about 35 per cent feldspar, which is seen as pinkish vitreous grains. The quartz grains are angular and have an iron oxide coating. Feldspar is readily weathered, especially in humid conditions, and a high percentage in a detrital rock, such as arkose, may indicate rapid deposition and/or an arid environment. (C)

Grit A coarse- or medium-grained quartz-rich sediment with angular grains, in this case deposited by water. Some mica and feldspar is present, but the latter mineral is by no means as abundant as in an arkose. This specimen is well sorted, with the grains mostly the same size, and there is a coating of yellowish iron oxide which acts as a poor cement. (CP)

Micaceous sandstone A specimen of fine-grained sandstone with much muscovite mica on the bedding planes. Mica flakes tend to be carried away by the wind in arid continental environments and their abundance in a sandstone is one factor which suggests water deposition. (CP)

Clay A very fine-grained sediment which even when examined microscopically reveals only a little detail. It is composed of clay minerals and very fine grains of quartz, mica and feldspar. When wet, clays become sticky as a thin film of water adheres to the multitude of flaky particles in the rock. (CP)

Greywacke This is a characteristically dark-coloured specimen of what is essentially a poorly sorted sandstone containing a variety of grain sizes from sand to clay. The chemistry of these rocks is somewhat complex with the larger angular fragments consisting of quartz and feldspar and rock fragments, while the finer matrix material contains clays, chlorite, quartz and pyrite. Such sediment may be deposited by rapidly moving turbidity currents, in deep ocean environments, but many of the sediments formed from such flows are graded. (K)

Mudstone is of the same grain size as shale but will not break as easily because of less well-defined bedding. The specimen shown is a non-marine mudstone with carbonized plant remains on the bedding surfaces. (CP)

Ripple-marked sandstone This is a reasonably fine-grained sandstone with some muscovite mica visible as small glinting flakes on the bedding surface. The well-preserved ripple marks are a common sedimentary structure which is useful in reconstructing the conditions of deposition. (CP)

The bedding surfaces of detrital sediments such as sandstones represent the sea bed or land surfaces of the past. These bedding planes may have preserved on them features indicative of the conditions under which they formed. Here two dipping sandstone bedding planes show such structures. The upper surface, to the left of the picture, shows desiccation cracks indicating that when forming this surface dried out and the resulting cracks were subsequently filled in with more muddy sand. The surface immediately below this has ripple marks on it indicating that shallow water moved over the sediment surface.

Sandstone in thin section

Chevron folds

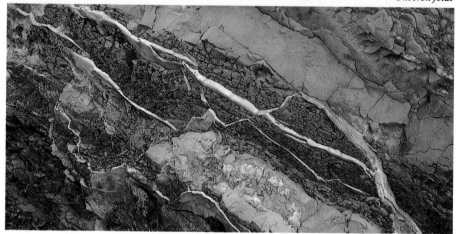

Stratified gypsum

Sandstone in thin section consists of quartz grains which do not always fit together well. Depending on their orientation these quartz grains are grey, white or black and have angular outlines. Quartz sand with grains of this size and shape is typically formed in river or marine environments. Small particles of brightly coloured mica and feldspar with 'striped' grains are also present. Some of the black grains are iron oxide.

Chevron folds On the north coast of Cornwall, UK, alternating sandstones and shales of the upper Carboniferous Crackington formation have been highly folded into these striking chevron folds. The development of such folds depends on the 'multi-layered' fabric of strata as in these turbidites with competent sandstones occurring with incompetent shales.

Evaporites

These are rather specialized rocks which need very particular conditions for their formation. They require the precipitation of the various salts from a saturated solution. Usually this occurs in arid climates. For example, when a lagoon or gulf becomes separated from the main body of the sea, by the lowering of sea-level or the deposition of a sand bar, the concentration of salts will increase with evaporation, and deposition of evaporites takes place. A similar situation can occur in an inland drainage basin, especially in arid regions, where salts brought in by river systems become concentrated. The minerals formed in each environment may vary. The marine evaporites, which are often interbedded with muds, marls, sands and dolomitic limestones, are mainly gypsum, anhydrite, halite and potash; those formed in non-marine situations are related to the minerals in the source area draining into the basin.

Evaporites are deposited in a strict sequence with the least soluble being precipitated first and most soluble last. Thus gypsum followed by rock salt and finally potash is a typical sequence. Many evaporite deposits have been exploited commercially, as the various minerals have specialized uses. Gypsum is used for making plaster and plaster boards, and both gypsum and anhydrite are used in the production of sulphuric acid. Rock salt is a preservative and is also used for making soap, insecticides and dyes. Potash is in great demand as an agricultural fertilizer. Non-marine evaporites include borax which can be used in enamel and glazes, and various nitrates used in fertilizers, nitric acid and explosives.

Rock salt This specimen is reasonably pure, being pale coloured. With an increase in impurities such as iron oxides and clay, rock salt becomes reddish-brown or dark grey. Thick beds of rock salt can become plastic under pressure at depth in the crust, and the salt flows upwards, disrupting overlaying strata to form domes and plugs. As well as being a source of rock salt, these plugs are of great interest as oil traps are frequently found against them. (CP)

Gypsum rock A specimen from a series of red sandstones and marls, among which this soft, often pink-coloured evaporite is frequently found. During compaction of sediments gypsum beds are easily distorted. (CP)

Marl with gypsum This is a typical association in evaporite beds and here thin gypsum layers are present along the bedding planes of the marl. (K)

Marl is a fine-grained detrital sediment similar to clay but it contains a high percentage of calcite. Marl can vary in colour from grey to green or red depending on impurities such as iron oxide. Both red and green marls are illustrated. (CP)

Potash ore, sylvinite Typically this is a greyish rock with patches of pale or reddish sylvite (sylvine) scattered through it. There is often much halite in the ore, the potash showing up as bright red patches when stained with hematite. When not stained with impurities the potash can be white. (C)

Stratified gypsum Gypsum may occur as strata as in this example where pale thin gypsum beds (which are about 3 cm thick) exist with beds of marl, clay and other evaporites.

Rock salt

Gypsum rock

Green marl

Marl with gypsum

Potash ore

Red marl

Potash ore

Calcareous mudstone

Crinoidal limestone

Dolomitized limestone

Pisolite

Organic limestone

Oolitic limestone

Dolomite

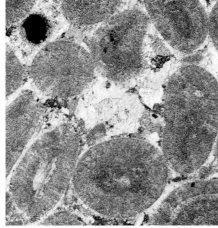

Oolitic limestone in thin section

Weathering of limestone

Limestones

Limestones or carbonate rocks are those which are essentially composed of carbonate minerals, usually calcite or dolomite. These rocks commonly contain a variety of other minerals in small amounts, including quartz and clay. Limestones can be deposited in both marine and fresh-water environments, the former being more common and often representing a moderately shallow, clear-water zone; the fresh-water limestones are recognized most readily by the fossils they contain. A broad classification of limestones generally includes three main divisions: organic limestones containing much fossil material, such as coral, crinoid fragments or algae; chemically formed limestones including oolites and dolomites; and clastic limestones, a category including mechanically formed carbonate rocks. All limestones are very susceptible to chemical weathering as the action of weak carbonic acid in rain-water (and many other acids in industrial areas) produces soluble bicarbonate of lime which is then carried away. As a result, a characteristic limestone landscape with much bare surface rock and underground caverns develops. Many limestones, because of the soils that develop on them and their drainage characteristics, support a unique flora containing many rare species. Limestone can have considerable economic importance, as a raw material for cement making, for example, or as a building stone.

Oolitic limestone in thin section At high magnification, the detailed structure of oolitic limestone can clearly be seen. In this example from the Jurassic the rounded and ovoid ooliths with a central nucleus are cemented by a mass of pale, brightly coloured calcite crystals. The concentric layering of calcite in the ooliths is especially clear in the bottom right of the picture.

Calcareous mudstone This is a very fine-grained sediment composed of calcite particles too small to detect with the naked eye. Some detrital material such as quartz and clay can be present and, on rare occasions, fossils. The typical sub-conchoidal fracture is well seen in this example. (K)

Crinoidal limestone An organic limestone which is composed almost entirely of the stems of crinoids and broken ossicles, held in a matrix of cleaved calcite fragments. Such rocks can be darker than this specimen when a higher proportion of detrital material is present. (CP)

Dolomitized limestone This specimen is of an originally calcite-rich limestone which has undergone replacement by dolomite. The greyish colouring is due to the detrital quartz and clay in the sediment. Dolomitization can be brought about by waters circulating through the sediment, and when this occurs fossils may be destroyed. (CP)

Pisolite The large rounded particles (pisoliths) are held in a calcareous matrix. They are formed by the precipitation of calcite around a nucleus, such as a shell fragment. Such chemical activity is helped by the constant agitation of warm, shallow sea water and the action of algae. (K)

Organic limestone In this specimen grey lime mud has cemented a mass of fossils together. They include rounded crinoid fragments, brachiopods, bryozoans and corals. (JHF)

Oolitic limestone This is commonly a pale-coloured rock varying from cream to yellowish or reddish. The ooliths are small (about 1 mm diameter), rounded or sub-rounded grains made of concentric layers of calcite precipitated around a nucleus in a similar way to the larger pisoliths. Constant agitation allows the even precipitation of the calcite from warm sea water. Many small fossil fragments can be seen in this specimen, mainly mollusc shells. Oolites are usually well-bedded, and structures like cross-bedding are common. Today, sediment of this type is forming in the warm, shallow, agitated waters around the Bahamas. (CP)

Dolomite These carbonate rocks contain well over 15 per cent magnesium carbonate and tend to be more creamy-brown than calcite-rich limestones, especially when weathered. The rock is associated with evaporite sequences. Small bivalve molluscs are fossilized in this example. (CP)

Weathering of limestone Because of their high calcium carbonate content, limestones are susceptible to chemical weathering, especially by rain water which is naturally weak carbonic acid. In industrial countries this rain water also contains many other acids including sulphuric and nitric. The joint systems common in many limestones are exploited by this water and they are enlarged as in this limestone pavement near Malham, Yorkshire, UK.

Freshwater limestone

Stalactite

Chalk

Calcareous tufa

Red chalk

Concretionary dolomite

Nummilitic limestone

Coquina

Coral limestone

Sea cliffs, Yorkshire, UK

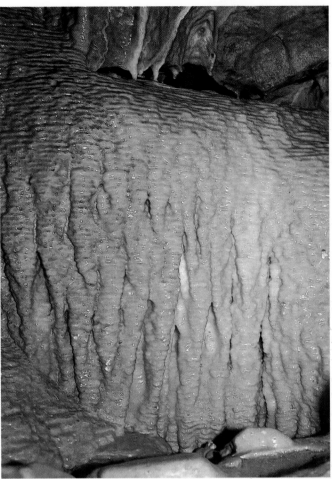

Limestone caves

Sea cliffs, Yorkshire, UK   The soft pure limestone called chalk forms high vertical sea cliffs in Britain and north-west Europe. Here the wave-cut platform is visible through the water, but as it is very soft the material eroded from the cliff face does not remain long enough to form a boulder beach. Cliffs like this are much favoured by nesting sea birds; the only mainland based colony of gannets in the UK is seen here.

Limestone caves   Cave systems develop along the much enlarged joint and bedding plane systems of many limestones. The water running and dripping through these caves is rich in bicarbonate of lime which is redeposited as calcareous tufa, in a variety of forms. Here a mass of stalactites hangs on a cave wall where a small stream issues.

Freshwater limestone   This is less common in the geological record than marine limestone and is usually recognized by the non-marine fossils it contains, in this case the gastropod *Viviparus*. Chemically, freshwater limestones are very similar to those formed in the sea. (K)

Chalk   is a very soft, often powdery, bio-micrite limestone of extreme purity. It is composed of innumerable micro-fossils, especially coccoliths and *Foraminifera*, set in a matrix of very fine lime mud. At least 90 per cent of the rock is calcium carbonate, and macro-fossils are not uncommon. The chalk found over much of western Europe is not unlike the deep sea muds (oozes) of modern oceans. It is thought, however, that the chalk accumulated in a sea some 200 or more metres deep, when sea-level over European continental shelves was much higher than today, and not in abyssal depths. The lack of detrital matter in the rock can be explained by suggesting that the neighbouring land areas were very low-lying and dry, with little erosion taking place. (CP)

Stalactite   In underground limestone caverns, caused by solution of the rock along joint systems and bedding planes, pendant accumulations of inorganically precipitated limestone tufa are called stalactites. Water seeping through the cavern roof and charged with lime deposits calcite when CO_2 is given off on contact with the air. Evaporation speeds the process. Stalactites are long slender forms, while stalagmites are short stumpy mounds growing up from the cave floor. (CP)

Calcareous tufa   This is the material, rich in calcite, which is deposited in limestone regions by precipitation from lime-rich waters. Tufa often coats cliffs, waterfalls and quarry faces, and can encrust mosses and other plants, preserving them in a crusty mass. When impurities such as iron are present, tufa can be yellow or red. If tufa binds a mass of pebbles, then calcrete is formed. Tufa is often non-bedded, slightly spongy and porous. A bedded variety of tufa which is common around hot springs is called Travertine. (K)

Red chalk   When a small percentage of ferric oxide is present, chalk takes on a brick-red colour. This specimen contains fossils, including a belemnite. (CP)

Concretionary dolomite   Concretionary mounds occur in many carbonate rocks, in this case in a buff-coloured dolomite. Such nodular masses may be the result of chemical segregations or algal growths; they are composed of the same material as the host rock. (CP)

Nummilitic limestone   An organic limestone made of the small, rounded shells of the foraminiferid, *Nummilites*. These are packed on to the bedding planes and cemented with calcite. (C)

Coquina   An unbedded mass of shell debris and limestone clasts cemented with calcite or dolomite. Such rocks are often associated with reef environments where sediment may accumulate by slumping or by the *in situ* growth of sediment bound by organisms like polyzoans. Coquina differs from other reef limestones in that it is essentially made of shell debris. (C)

Coral limestone   Many organic limestones contain abundant coral which is virtually in its position of growth, with lime mud cementing the corallites together. Coral-rich limestones are generally pure, containing little detrital sand or mud, as corals require clean sea water. (CP)

Banded ironstone

Coal

Chert

Chert in limestone

Flint

Ironstone

Ironstone nodules

Amber

Septarian concretion

Nodules and concretions

Coal, Ironstone, Concretions

Banded ironstone This specimen shows alternating layers of dark chert and red iron minerals, mainly hematite. Magnetite, siderite, pyrite and greenalite are also found in these rocks. Banded ironstones are known from Pre-Cambrian formations in many parts of the world including Australia, northern Europe, India, Brazil and Africa; this specimen is from Ontario, Canada. The vast majority of banded ironstones are restricted to the Pre-Cambrian, being between 2000 and 3000 million years old. They differ both in composition and distribution from younger iron-bearing sediments. Though the formation of these rocks, which contain iron oxides produced at a time in the past when oxygen was a rare gas in the atmosphere, is somewhat problematical, it is generally believed that they were deposited as precipitates (possibly aided by bacterial action) in partially enclosed basins or in large lakes. The high levels of carbon dioxide in the atmosphere when these rocks formed would have produced very acid rain and river water capable of carrying iron compounds in solution. (K)

Coal This is an example of a high-ranking, relatively hard, vitreous coal. Such coal has a high carbon content and low volatile percentage. Less pure coals are dirty to the touch and have a dull lustre. Essentially, coal is formed from vegetable peat through the increase of pressure and especially heat due to burial and compaction. The heat produced drives off volatiles, and coals near to intrusive igneous masses may be of exceptionally high rank. Coals show bedding structures and may contain plant fossils indicating their sedimentary origin. There is a marked decrease in the quality of coal in formations which are geologically young. Some Jurassic coals are of good quality but many Tertiary coals are brown lignites. The best quality coals are of Carboniferous age. (CP)

Chert is a very hard rock material found as discrete masses in limestone. It is usually dark in colour and of cryptocrystalline structure. Its origin is open to some debate. It may be formed from organic silica, for example the remains of sponges or microfossils. The accumulation of the silica occurs in colloidal form on the sea bed. Some chert deposits are found associated with deep-water pillow lavas. (K)

Chert in limestone This specimen of dark Cambrian limestone contains small rounded patches of chert. Being siliceous, chert is much harder than the limestone and also does not react with cold dilute hydrochloric acid. (CP)

Flint This form of chert occurs as bands of nodules in the Cretaceous chalk. It is commonly black or dark grey in colour and individual flints which have been weathered from the chalk often have a chalky coating. Flint breaks with a very sharp conchoidal fracture and may contain fossils, often of sponges or echinoids. (CP)

Ironstone This specimen is of oolitic ironstone from Normandy, France. Iron oxide has been precipitated from solution, possibly below the surface of the sediment. Such oolitic

Chert in submarine lavas

Banded ironstone from the Yilgarn block, Australia

ironstone has all the characteristics of carbonate sediment, including fossils and bedding structures, but iron oxide takes the place of calcium carbonate. (K)

Amber A soft resinous to sub-vitreous material with a conchoidal fracture, amber is formed from hardened coniferous tree resin. It sometimes contains fossilized insects which became stuck in the viscous resin as it oozed from the tree trunk. It has frequently been used for jewellery making and can be found on the shores of northern Europe, where it has been weathered from Tertiary and Pleistocene deposits. (CP)

Ironstone nodules Rounded and ovoid nodules are quite frequently found in shales and clays. They are commonly composed of pyrite or siderite and may occur in rows following the sedimentary bedding planes. The precipitation of the nodule material may be initiated around a fossil and perfect specimens are sometimes contained. (CP)

Septarian concretion As with the formation of iron-rich nodules, other minerals can become

segregated during diagenesis and may form distinct rounded masses. In this example (which has been sliced to show the internal structure of septa) a calcareous concretion has developed in pelitic sediment and shrinkage cracks have been infilled with white calcite. (C)

Nodules and concretions are common along bedding planes of shales and clays. Here large ovoid concretions about 30 cm in diameter have been exposed by marine erosion. Being more resistant than the shale, the concretions protect small areas of sediment and stand above the level of the wave-cut platform.

Chert in submarine lavas The margins of two rounded masses of pillow lava, with vesicular texture, occupy the sides of this picture, while the centre is a mass of greyish chert filling the gap between the pillows. This specimen is at Pentire Head, Cornwall, UK.

Banded ironstone from the Yilgarn block, Australia This specimen shows the dark iron-rich layers alternating with pinkish quartz bands. The specimen is 10 cm across. (AF)

Minerals

The rocks which make up the Earth's crust are composed of combinations of minerals. These vary tremendously in their chemical composition and physical properties. A number of minerals are of great economic importance, for example the ores of metals; other minerals, because of their great beauty, or the perfection of their crystal form, are highly prized as gems. Despite the great variety of minerals, any individual mineral will have physical and chemical properties which vary only within known limits. In order to identify a given mineral a number of easily determined properties can be examined.

Crystals and Habit

One of the most apparent features of a mineral is its shape. Depending on the combination of the atoms which form a mineral so the crystal shape will develop. Crystals of the same mineral will have similar shapes; for example, fluorite commonly forms as cube-shaped crystals and quartz crystallizes as hexagonal prisms. The plane surfaces which bound the crystals reflect the internal atomic arrangement. Such well-formed crystals are only able to grow where there are no confining limits. Depending on how the plane faces develop and what combinations occur, a great variety of crystals can be formed. In order to understand and classify crystals, their symmetry is considered, and a number of crystal systems based on their degree of symmetry are used. A solid crystalline shape, such as a cube, can have a number of axes and planes of symmetry. An axis of symmetry is an imaginary line through the centre of the solid, round which it can be rotated to give the same view of the solid more than once; a plane of symmetry is an imaginary surface along which the solid could be cut to make a 'mirror image'. Our example, the cube, has a great many axes and planes of symmetry and from these only a few are chosen in order to define the cubic crystal system. The axes of symmetry which are selected are called the crystallographic axes. Those selected for the cubic system are the ones which cut the centres of the opposite faces of the cube, and the cubic system is defined as having three axes of equal length which intersect at right angles. Many solid shapes besides the cube fit this definition. The octahedron, rhombdodecahedron and trapezohedron all fall within the cubic system. In all there are six crystal systems based on crystal symmetry: the Cubic, Tetragonal, Orthorhombic, Monoclinic, Triclinic and Hexagonal systems. Some experts also define a seventh system, the Trigonal, which is very similar to the Hexagonal. Each system contains many possible shapes. Crystal forms often enclose space, for example the cube, but other forms like pinacoids (two parallel faces) and prisms (a number of faces meeting at their parallel edges) are open forms and can only exist in combination with other faces.

The term habit is given to the actual shape that the specimen of a mineral exhibits and for identification purposes this is probably the most important aspect of mineral shape. Habit can vary from a fine crystal which can be readily placed in one of the systems to an amorphous mass with no definite shape. There are a number of words used commonly to describe habit. These include:

Tabular a generally flat shape with broad faces.
Reniform a rounded shape, like a kidney.
Fibrous very thin prismatic crystals, often thread-like.
Botryoidal like a bunch of grapes.
Mamillated large, gently rounded aggregates.
Granular a mass of grains.
Acicular many needle-shaped crystals, often radiating.
Massive of no distinct shape.
Dendritic with tree-like branching shapes.

Colour and Streak

Among the whole range of minerals there is great variety of colour which in some cases is of diagnostic value. For example, malachite is a rich green, galena is lead grey, amethyst quartz is purple and iron pyrites (fool's gold) is a silvery-gold colour. The problem with colour as a means of identification is that many minerals are the same colour, while others exhibit a variety of colours. A good example of the latter is quartz, a very common mineral, which can be clear and colourless, white, green, purple, pink, brown or black. The colour varieties of quartz are even given different names. The colour can be a result of impurities in the crystal lattice or of the way light falling on to the crystal is broken up by absorption or refraction.

Streak is the colour of the powder of the mineral, usually obtained by drawing the specimen across an unglazed porcelain streak-plate. Hard minerals may have to be crushed or scratched with a harder object. Even though some minerals like quartz have a variety of colours, their streak is constant, quartz having a white streak.

Lustre

The way in which a mineral's surface reflects light produces a variety of different lustres. Some minerals, such as pyrite and galena, have a metallic sheen or lustre, whereas the glassy surface of many minerals like quartz and fluorite is said to have a vitreous lustre. Other lustres may be described as resinous, pearly, silky, and earthy. Diamond has a brilliant adamantine lustre.

Cleavage and Fracture

Not only can crystals become damaged by weathering and erosion, but they may be easily broken and fractured or cleaved. Fracture is the irregular breakage which produces uneven surfaces that are in no obvious way related to atomic structure of the mineral. All minerals fracture and some produce a shell-like curved, conchoidal fracture, rather like broken glass. Cleavage, on the other hand, is definitely related to the bonds between the atoms of the mineral. The breakage occurs where the bonds are weakest and cleavage surfaces are relatively smooth planes which reflect light evenly and can be produced many times over for a given mineral. The cleavage fragments will be a characteristic shape, like the rhombs of calcite, or flakes of mica.

Hardness

This is the resistance of the surface of a mineral to being scratched. A ten-point scale with the intervals represented by well-known minerals is used to judge this property. This was established in 1812 by F. Mohs, and that it has remained a good yardstick for so long is evidence of its effectiveness. The scale is as follows:

1. Talc	6. Orthoclase
2. Gypsum	7. Quartz
3. Calcite	8. Topaz
4. Fluorite	9. Corundum
5. Apatite	10. Diamond

The intervals on this scale are not of equal hardness; diamond is many times harder in real terms than corundum, and talc and gypsum are only very slightly different. In order to test this important property minerals from the scale can be used, but some other common objects are also handy. A finger nail is 2½ on the scale of hardness, a coin is about 3½ and a steel knife blade is 5½. When testing hardness by rubbing two minerals together care must be taken as it is not always easy to detect which mineral is being marked.

A volcanic crater in north-west Hokkaido, Japan

A volcanic crater in north-west Hokkaido, Japan This crater is yellow with encrusting native sulphur deposited from issuing gases. This is a common way in which sulphur deposits occur, some of which are large enough to be of economic use.

Twinning

Twinned crystals consist of two parts which are orientated in a different way, but which share a crystallographic surface. This surface, the twin-plane, is a possible crystal face. Crystal development takes place from a sheet of atoms and so growth may occur on both sides but the crystal shapes thus formed will have a different orientation. Twin planes can be planes of symmetry, in which case the crystals on each side of the plane will be mirror images. Many different types of twinning are recognized, including simple and polysynthetic, repeated, complex and penetration twins.

Specific Gravity (Relative Density)

An obvious feature of many mineral specimens is that though they may be roughly the same size, their weight is very different. Equal-sized specimens of galena (lead sulphide) and halite (sodium chloride) would be very different, the galena being far heavier. The atoms of some elements are larger than others and they may be more closely packed together. Specific gravity (SG) is a scientific way of measuring this obvious weight difference by comparing objects against a standard, usually water. The weight of the mineral is divided by the weight of an equal volume of water (obtained by displacement) and the figure thus obtained is the specific gravity. Galena, mentioned above, has a specific gravity of 7.5 and halite is 2.2. Common quartz is 2.65 and gold is about 19. With practice it is possible to give a reasonable estimate and at least say whether the specimen being considered is greater than quartz or other common specimens.

Other Properties

The properties explained briefly above are those commonly used for mineral identification, and

Carrock mine, Cumbria, UK

Mineral colour and habit

Mineral colour and habit

are those which do not require any special equipment. In the section of mineral illustrations these properties are given for all the minerals described. There are, however, a number of minerals which exhibit other properties, which may be special to the mineral. Where this is the case it is indicated in the captions. Such properties include magnetism, reaction with weak acids, solubility in water and fluorescence and radio-activity.

Mineral Formation and Occurrence

Minerals form as important components of rocks, and develop in cavities and other suitable sites in the crust. The rock-formers are mainly silicates, which develop from magma or lava. Good crystals may not develop because of the confines of the rock matrix, but where drusy cavities occur, as in granites, fine quartz and feldspar crystals can be found. The main rock-forming silicates are feldspars, quartz, micas, pyroxenes, amphiboles and olivines. The minerals formed in metamorphic rocks are in many cases similar to the igneous silicates and include silicates of aluminium, kyanite, sillimanite, cordierite and andalusite along with minerals such as garnet, serpentine and talc. Sedimentary rocks are often derived from other materials and include the resistant minerals from preformed rocks along with some new specifically sedimentary minerals. Quartz and feldspar are common with new clay minerals developed from the alteration of, for example, feldspars. Calcite and dolomite are rock-formers in limestones, and evaporites like gypsum and rock salt (halite) are special to certain sedimentary environments.

The finest mineral specimens are those found in non-rock-forming situations, apart from the giant crystals which may form in some pegmatite veins. Many of the greatest concentrations of minerals, especially metallic ores, are formed by hydrothermal processes. Such minerals occur in veins and fractures cutting through strata. They are precipitated from migrating fluids which are hot and chemically very active. The source of the fluids can vary. They may be residual fluids from magmatic cooling, or may be developed from deep brines trapped in sediments. Metallic ores of lead, copper and zinc are common in veins deposited from these fluids, along with tin and iron. As well as the ore minerals, many other 'gangue' minerals occur in hydrothermal veins. These include quartz, calcite, fluorite, barite and siderite. The gangue is the part of the vein infilling which at the time of ore extraction is of no economic worth. When economic conditions change it may be that a previously worthless gangue may become valuable. The minerals in a hydrothermal field may be deposited zonally, even with a concentric organization around an igneous body. Some metallic orefields have a zone of tin ore nearest to a granite body, with copper, zinc and lead in zones out from the intrusion. Pneumatolysis is a process related to hydrothermal activity. This brings about changes by the action of gases such as fluorine and boron fluorides, at a late stage in the cooling of a granitic or similar body. Tourmaline is a common mineral produced in this way, and wolfram and tin are two important ores thought to be formed by pneumatolysis. Pegmatites, which contain spectacular large crystals, may be formed during the late stages of magma consolidation by emanating fluids. Exceptional crystals of quartz, feldspars and mica together with beryl, topaz, apatite and uranium minerals

'Artist's Palette', Death Valley, California

occur. Fluids deep in the crust can produce replacement of rocks by minerals. Limestone can, for example, be replaced by hematite over wide areas.

Volcanic eruptions can produce a variety of minerals, some of economic significance. Around vents accumulations of sulphur and related minerals containing antimony and mercury occur, but far richer mineral deposits are found near submarine vents. Manganese, zinc, copper and iron-rich brines exist in such situations, for example deep in the Red Sea. Deposits rich in iron and manganese are found near some ocean ridge systems, possibly resulting from a reaction between the volcanic rocks and sea water. Interpretation of the theories concerned with plate tectonics shows how metal-rich ore deposits are related to ancient plate boundaries. The rich copper ores in Cyprus are associated with ocean crust formation and the ore fields of the Andes and the Western Cordillera of North America are closely related to subduction zones. Some of the richest copper deposits, porphyry copper ores, are formed in sub-volcanic environments, and many of these are situated along the west coast of South and North America. As magma rises, volatiles generate such pressure that the country rock becomes shattered, and minerals are deposited in this resulting vein system. Such deposits are often associated with porphyritic granodiorites intruded as stocks at a depth of only a few kilometres.

Other extensive formations of nickel, chromium and platinum are found in large basic and ultra-basic igneous intrusions. These metal ores accumulate through magmatic differentiation. They sink through the liquid magma because of their high specific gravity, to become concentrated low down in the intrusion, often with early formed silicates like olivine and pyroxene. The Bushveldt intrusion in South Africa is a series of lopoliths of Pre-Cambrian

age, made of gabbro and anorthosite, with a total thickness of about 7 km. Layered accumulations of chromium, platinum and iron ores occur here.

Sediments are generally not as rich in minerals as igneous or metamorphic rocks, but where placers occur great concentrations of certain, often valuable, ores exist. Placers are accumulations of heavy or resistant minerals weathered from already formed rocks and sorted by running water or tidal currents. Such deposits can occur on beaches and in the sands and silts of river basins. Beaches may be black with concentrations of magnetite, ilmenite or chromite. Others minerals which form placers are gold, cassiterite, uranium minerals, copper ores and some gem-stones such as diamond, ruby and sapphire. Areas where placers are important include the Witwatersrand conglomerates in South Africa, where Pre-Cambrian sediments with gold and uranium occur. In Malaysia and Thailand there are unconsolidated sands rich in tin, and the Jacobina placers in Brazil contain gold. Conglomerates rich in uranium placers occur near Johannesburg, while the central African copper belt is formed from modified placers where copper minerals are found in shales around algal reefs.

Weathering can produce other types of mineralization as well as providing the basic material of placers. Metallic sulphides which have accumulated in veins are susceptible to chemical weathering, and the metals thus made available can be carried down to deeper levels in the vein system to be re-deposited in a zone of secondary enrichment. Copper, lead, zinc and iron minerals can occur like this. The weathered surface zone, often made of brownish-yellow iron hydroxides, is called 'gossan'; its presence indicates to prospectors the possibility of rich ore lodes deeper in the vein. It is also possible for some metals which are not readily carried in

solutions to become concentrated as residual deposits in the weathered mantle. Chemical weathering can be very active in tropical regions with a distinct wet season. Bauxite rich in aluminium and iron-rich laterites are formed in such areas. If there is a distinct dry and wet season, leaching will take place in the wet season, removing soluble materials. In the dry season capilliary activity brings salts to the surface; these are washed away by the next rains. This chemical system produces concentrations of iron and aluminium oxides, depending on the parent rock. Basalts tend to produce laterite and bauxite is found above granites.

Metamorphic rocks contain a variety of minerals related to the often extreme conditions of temperature and pressure to which they have been subjected. As has been said, many of these are silicates not unlike those found in igneous rocks, but talc, asbestos and graphite are also commonly found in metamorphic rocks.

'Artist's Palette', Death Valley, California
Sedimentary rocks are frequently brightly coloured by cementing minerals. Here in Death Valley, California, an arid region influenced by periodic surface water, copper minerals colour the strata of the aptly named 'Artist's Palette'.

Mineral colour and habit These two properties are well illustrated by this green hexagonal prism of beryl and the contrasting silvery radiating specimen of marcasite.

Carrock mine, Cumbria, UK, is at the head of a narrow road leading among the fells. In the early years of this century the exploitation of tungsten minerals began here. The subsequent history of mining in this remote region has been much influenced by war and the world market, and some very recent exploitation has taken place.

Sulphur

Sulphur

Sulphur

Arsenic

Silver

Copper

Silver ore

Copper

Graphite

Antimony

Silver

Graphite

Diamond

Bismuth

Gold

Diamond *Chemistry* C. *System* Cubic. *Habit* rounded octahedral crystals. *Colour* colourless, black, grey, green, yellow, brownish. *Streak* white. *Cleavage* octahedral. *Fracture* conchoidal. *Hardness* 10. *SG* 3.5. *Lustre* adamantine. *Special features* hardness. *Formation* in ultra-basic rocks called kimberlites which form pipe-shaped intrusions from great depth; also in detrital deposits derived from kimberlites. *Distribution* South Africa, Arkansas (USA), Yakutia (USSR), Brazil, Venezuela, Ghana. (L)

Sulphur *Chemistry* S. *System* Orthorhombic. *Habit* tabular crystals, massive. *Colour* yellow. *Streak* white. *Cleavage* none. *Fracture* uneven. *Hardness* 1½–2½. *SG* 2.1. *Lustre* resinous. *Special features* colour, low SG, insoluble in water and weak acids, soluble in carbon disulphide. *Formation* encrusting the crevices and craters of volcanoes and hot springs. In the upper parts of salt domes. *Distribution* Sicily, Italy; Utah, California and Texas (USA) and Japan. (K, LW)

Silver *Chemistry* Ag. *System* Cubic. *Habit* wires and scales; crystals rare. *Colour* silver-white, tarnishes to brown, yellow and grey. *Streak* shining silver-white. *Cleavage* none. *Fracture* irregular and rough, malleable. *Hardness* 2½–3. *SG* 9.6–11. *Lustre* metallic. *Special features* high SG, colour, soluble in nitric acid, malleable. *Formation* hydrothermal veins with sulphides, in the oxidized zones of ore deposits and in placers with gold. *Distribution* worldwide, especially Kongsberg (Norway), Michigan and Nevada (USA), Broken Hill (Australia), Batopilas and Chihuahua (Mexico), Harz and Freiberg (Germany), Cobalt (Ontario, Canada), the Mendips, Pennine ore field and Leadhills (UK). (K, S)

Copper *Chemistry* Cu. *System* Cubic. *Habit* in branching and dendritic shapes; crystals can be rhombdodecahedra and cubes. *Colour* brown to coppery red. *Streak* red-brown. *Cleavage* none. *Fracture* rough. *Hardness* 2½–3. *SG* 8.9. *Lustre* metallic. *Special features* soluble in nitric acid, malleable, colour. *Formation* as a replacement mineral in sandstone conglomerates and other detrital sediments. In cavities in basalts and in the alteration zones of sulphide deposits and the weathered parts of copper lodes. *Distribution*

Lake Superior (Canada), Keeweenaw and Bingham (USA), Kurkoro (Japan), Kambalda (Australia), Bogoslovsk and Turinsk (USSR), Chile, Peru, Zimbabwe, Katanga. Cornwall and the Lake District are important in the UK. (K, L)

Arsenic *Chemistry* As. *System* Trigonal. *Habit* massive, reniform, granular, rarely as crystals. *Colour* greyish-brown. *Streak* pale grey. *Cleavage* perfect basal. *Fracture* uneven. *Hardness* 3½. *SG* 5.7. *Lustre* submetallic. *Special features* when heated smells of garlic. *Formation* minor constituent of hydrothermal veins with lead, nickel and silver. Associated with igneous and metamorphic rocks. *Distribution* the USA, Canada, Mexico, Sweden, France and Belgium. (K)

Antimony *Chemistry* Sb. *System* Hexagonal. *Habit* crystals rare, lamellar, granular, massive. *Colour* pale grey. *Streak* grey. *Cleavage* perfect basal. *Fracture* very brittle, uneven. *Hardness* 3–3½. *SG* 6.6–6.7. *Lustre* metallic. *Formation* hydrothermal veins with silver and arsenic. *Distribution* South Africa, China, Australia, Mexico, Turkey and Yugoslavia. (K)

Graphite *Chemistry* C. *System* Hexagonal. *Habit* massive, granular, tabular crystals rare. *Colour* black. *Streak* black. *Cleavage* perfect basal. *Hardness* 1–2. *SG* 2.2. *Lustre* metallic or dull. *Special features* low hardness, colour, greasy feel. *Formation* in schists, slate and marble, as flakes. In pegmatites with feldspar and quartz; also in veins. *Distribution* Sri Lanka

(22 cm-wide veins), Ontario, Korea, Bavaria, Essex County (New York, USA), Borrowdale (Cumbria, UK). (CP)

Gold *Chemistry* Au. *System* Cubic. *Habit* crystals rare, usually as grains, dendritic shapes and rounded nuggets. *Colour* yellowish-golden; with increasing silver content becomes much paler. *Streak* yellow-gold. *Cleavage* none. *Fracture* rough, malleable. *Hardness* 2½–3. *SG* 15.5–19.3, depending on purity. *Lustre* metallic. *Special features* colour, very high SG, non-tarnishing, malleable, insoluble in single acids. *Formation* often with quartz (as here) in hydrothermal veins. Also frequently in placers and consolidated sands and conglomerates. *Distribution* great amounts from Rand, South Africa. Also from India, Brazil, Bolivia, California and Alaska (USA), Mexico, New South Wales and Queensland (Australia), Austria, the Urals (USSR), Leadhills (S. Scotland), North Wales, E. Sutherland and Cornwall (UK). (L)

Bismuth *Chemistry* Bi. *System* Trigonal. *Habit* dendritic masses and massive. *Colour* silvery to pinkish when tarnished. *Streak* silver-white. *Cleavage* basal. *Fracture* uneven, brittle. *Hardness* 2–2½. *SG* 9.7–9.8. *Lustre* metallic. *Special features* colour and density. Fuses at 270°C. *Formation* with nickel, cobalt, silver and tin in hydrothermal veins. *Distribution* Kongsberg (Norway), Erzgebirge (GDR), Sardinia, Italy, Malaya, Bolivia, Australia, Colorado and Connecticut (USA), Canada, Cornwall (UK). (C)

Sphalerite

Orpiment

Molybdenite

Sphalerite

Covellite

Stibnite

Sphalerite

Sphalerite

Stibnite

Bornite

Chalcopyrite

Chalcopyrite

Molybdenite

Chalcosine

Enargite

Chalcosine (Chalcosite) *Chemistry* Cu_2S.
System Orthorhombic. *Habit* commonly
massive, or granular; pseudo-hexagonal or
tabular crystals are rare. *Colour* dull dark grey.
Streak black. *Cleavage* poorly exhibited
prismatic. *Fracture* conchoidal. *Hardness* 2½–3.
SG 5.5–5.8. *Lustre* metallic. *Special features* a
dark opaque mineral which dissolves readily in
nitric acid. *Formation* associated with
hydrothermal sulphides, usually in the zone of
secondary enrichment. *Distribution*
Connecticut, Miami, Montana and Arizona
(USA), Tsumeb (Namibia), Chile, Peru,
Mexico, the USSR, Calabona (Italy),
Frankenberg (Germany), Transvaal (S. Africa)
and Cornwall (UK). (S)

Sphalerite *Chemistry* Zns. *System* Cubic.
Habit tetrahedra and rhombdodecahedra,
crystal edges often curved. *Colour* yellowish-
brown to black. *Streak* brownish-yellow,
brown. *Cleavage* perfect, parallel to
rhombdodecahedron. *Fracture* conchoidal, very
brittle. *Hardness* 3½–4. *SG* 3.9–4.2. *Lustre*
resinous, can be adamantine. *Special features*
quite high SG, very brittle, colour variation
causes problems. *Formation* usually in
hydrothermal veins in association with many
other minerals such as galena, baryte, fluorite
and chalcopyrite. Also as a replacement
mineral, with magnetite and pyrite. *Distribution*
Missouri, Oklahoma and Kansas (USA), Oruro
and Potosi (Bolivia), Yugoslavia,
Czechoslovakia, Hungary, Italy, Switzerland,
Broken Hill (Australia); Cumbria, North
Pennines, Durham, Cornwall and Derbyshire
(UK). (BS, JM^C, CP)

Chalcopyrite *Chemistry* $CuFeS_2$. *System*
Tetragonal. *Habit* massive or granular, pseudo-
tetrahedral crystals not common. *Colour* brassy-
yellow. *Streak* greenish-black. *Cleavage* poor.
Fracture uneven. *Hardness* 3½–4. *SG* 4.1–4.3.
Lustre metallic. *Special features* soluble in nitric
acid, much yellower than iron pyrites and far
less hard. *Formation* in hydrothermal veins, in
porphyry copper deposits, as veins in diorites
and in thermally metamorphosed rocks.

Distribution Utah, Arizona and Montana (USA);
the Urals (USSR), Chile, Sudbury (Canada),
France, Sweden, Yugoslavia, Rio Tinto (Spain),
Germany, Italy, Cyprus and South Australia;
Cornwall, Cumbria, Wanlockhead and the
Pennines (UK). (K, CP)

Bornite *Chemistry* Cu_5FeS_4. *System* Cubic.
Habit commonly massive, cubes or octahedra.
Colour copper-red but tarnishes readily to
iridescent purple called Peacock Ore. *Streak*
greyish-black. *Cleavage* none. *Fracture* uneven.
Hardness 3. *SG* 4.9–5.4. *Lustre* metallic. *Special
features* iridescence, soluble in nitric acid, high
SG. *Formation* in hydrothermal veins and from
magmatic segregation. *Distribution* Connecticut
and Montana (USA), Chile, Mexico, Peru,
Australia, Zambia, Zermatt (Switzerland),
Gross Venediger (Austria), Cornwall and
Cumbria (UK). (JHF)

Covellite (Covelline) *Chemistry* CuS. *System*
Hexagonal. *Habit* massive or tabular crystals
with a platy form. *Colour* deep indigo blue, with
some iridescence. *Streak* grey-black. *Cleavage*
perfect basal. *Fracture* uneven. *Hardness* 1½–2.
SG 4.6–4.8. *Lustre* metallic to dull. *Special
features* a very soft mineral with perfect cleavage
and distinctive colour. *Formation* in
hydrothermal veins and in zones of secondary
enrichment of copper lodes. *Distribution* Butte
(Montana, USA), Alaska, Bolivia, Chile,
Sardinia, Italy, Yugoslavia, and many other
regions with copper veins. (K)

Molybdenite *Chemistry* MoS_2. *System*
Hexagonal. *Habit* usually as scaly masses; may
be granular and in hexagonal plates. *Colour*
lead-grey with a bluish tint. *Streak* blue-grey.
Cleavage flexible non-elastic laminae are
produced by perfect basal cleavage. *Hardness*
1–1½. *SG* 4.6–4.8. *Lustre* metallic. *Special
features* higher SG than graphite, greasy to
touch, perfect cleavage, very soft. *Formation* a
mineral which is never found in large quantities,
it occurs in granites as an accessory mineral and
in quartz and pegmatite veins. *Distribution* New
Jersey and Colorado (USA), Mexico, Bolivia,

Australia, Mulbach (Austria), Dunje
(Yugoslavia), Czechoslovakia, Italy, Norway,
Cornwall and Cumbria (UK). (K, CP)

Stibnite *Chemistry* Sb_2S_3. *System*
Orthorhombic. *Habit* acicular and prismatic
crystals which are striated parallel to their long
axis. Sometimes as massive or granular pieces.
Colour lead-grey. *Streak* lead-grey. *Cleavage*
perfect, parallel to length of the crystal. *Fracture*
brittle, sub-conchoidal. *Hardness* 2. *SG* 4.5–4.6.
Lustre metallic, dull when tarnished. *Special
features* melts in the flame of a match, very soft,
perfect cleavage. *Formation* in hydrothermal
veins with quartz and in deposits from hot
mineral springs. *Distribution* Shikoku (Japan),
Romania, Yugoslavia, Czechoslovakia, Hunan
(China), California (USA), Peru, Mexico,
Bolivia, Italy, and Cornwall (UK). (K)

Orpiment *Chemistry* As_2S_3. *System*
Monoclinic. *Habit* rare in crystals, usually
massive or foliated. *Colour* golden yellow,
brown and orange. *Streak* yellow. *Cleavage* one,
perfect, producing thin flexible laminae.
Hardness 1½–2. *SG* 3.5. *Lustre* pearly on
cleavage planes and otherwise resinous. *Special
features* colour is very distinctive, very soft.
Formation in low temperature mineral veins,
often with realgar; also deposited from hot
springs. *Distribution* Georgia (USSR),
Kurdistan (Turkey), Iran, Nevada (USA) and
the fumaroles of Vesuvius (Italy). (K)

Enargite *Chemistry* Cu_3AsS_4. *System*
Orthorhombic. *Habit* tabular crystals, lamellar
or granular, massive. *Colour* black. *Streak*
black. *Cleavage* prismatic and pinacoidal.
Fracture uneven. *Hardness* 3. *SG* 4.4. *Lustre*
metallic. *Special features* cleavages; melts in a
candle flame. *Formation* in low to medium
temperature hydrothermal veins, associated
with pyrite, tetrahedrite, sphalerite, bornite and
chalcosine. *Distribution* Chile, Peru, Mexico,
Argentina, Phillipines; Utah, Montana and
Colorado (USA), Italy, Yugoslavia and
Namibia. (C)

Tetrahedrite

Stannite

Pyrite

Arsenopyrite

Marcasite

Pyrite

Marcasite

Cobaltite

Tetrahedrite

Bournonite

Pyrite (Iron Pyrites) *Chemistry* FeS$_2$. *System* Cubic. *Habit* cubes, octahedra, pyritohedra, massive and in nodules. The faces of the cubes are frequently striated. *Colour* brassy-yellow, often quite pale. *Streak* greenish-black. *Cleavage* very indistinct, cubic. *Fracture* uneven, conchoidal, brittle. *Hardness* 6–6½. *SG* 4.8–5.2. *Lustre* metallic. *Special features* colour, striated cubes, harder than chalcopyrite and paler coloured. *Formation* in a very wide range of geological situations, including hydrothermal veins, as an accessory in igneous rocks, in slates and dark shales where it often replaces fossils and forms nodules. *Distribution* a very common mineral with a wide distribution. The biggest deposits are at Rio Tinto in SW Spain, but large bodies are also found in Tasmania and the Harz mountains in Germany, and in Japan; Colorado, Arizona and Pennsylvania (USA), Sulitjelms (Norway), Canada, South Africa and Italy. (JHF, CP)

Marcasite *Chemistry* FeS$_2$. *System* Orthorhombic. *Habit* flat prismatic crystals often as cockscombs, nodular and radiating. *Colour* paler bronze-yellow than pyrite. *Streak* dark grey. *Cleavage* poor, prismatic. *Fracture* uneven. *Hardness* 6–6½. *SG* 4.9. *Lustre* metallic. *Special features* paler than pyrite, with a slightly lower SG, 'cockscomb' habit. *Formation* in hydrothermal veins at lower temperatures than pyrite. Also found in sediments such as chalk, where it occurs as nodules with a radiating internal structure. *Distribution* common in many areas, Germany in veins; with lignite in clays in Czechoslovakia; Missouri and Illinois (USA), Romania, and British chalk. (K)

Arsenopyrite *Chemistry* FeAsS. *System* Monoclinic. *Habit* elongated prismatic crystals, pseudo-orthorhombic, granular and massive. *Colour* silver-grey with a brown tarnish. *Streak*

dark grey to black. *Cleavage* prismatic. *Fracture* uneven. *Hardness* 5½–6. *SG* 5.9–6.2. *Lustre* metallic. *Special features* colour, high SG, form of crystals. *Formation* in veins of hydrothermal origin often with gold, silver, tungsten and tin. Also in pegmatites and metamorphic deposits. *Distribution* Freiberg (GDR), Boliden (Sweden), Sulitjelma (Norway), Deloro (Canada), Yugoslavia, Italy and Austria. Mexico, Bolivia, Colorado (USA) and Cornwall (UK) have produced fine crystals. (K)

Stannite *Chemistry* Cu$_2$SnFeS$_4$. *System* Tetragonal. *Habit* massive or granular, crystals rare. *Colour* steel-grey to bronze. *Streak* black. *Fracture* uneven, very brittle. *Hardness* 4. *SG* 4.4. *Lustre* metallic. *Special features* colour and fairly low hardness, high SG. *Formation* in hydrothermal veins with galena, cassiterite, sphalerite and chalcopyrite, also associated with tin deposits. *Distribution* in many famous tin regions such as Malaysia, Sumatra (Indonesia), China and Bolivia; also in Cornwall (UK). (JM^C)

Tetrahedrite-Tennantite group *Chemistry* (Cu,Fe)$_{12}$ (Sb,As)$_4$S$_{13}$. *System* Cubic. *Habit* massive, granular, crystals tetrahedral. *Colour* tetrahedrite is steel-grey, tennantite tends to be bluish-grey. *Streak* dark grey to black. *Cleavage* none. *Fracture* uneven to subconchoidal. *Hardness* 3–4½. *SG* 4.6–5.2. *Lustre* metallic to dull. *Special features* dark colour, fairly soft, tetrahedral habit. *Formation* in hydrothermal veins usually with copper, silver, lead and zinc ores. Tennantite is more usually found in metasomatic bodies in limestones. *Distribution* in many areas in association with copper, lead and zinc deposits, for example in Nevada, New Mexico, California and Arizona (USA), Bolivia, Chile, Peru, British Columbia (Canada), Boliden (Sweden), Pribram (Czechoslovakia), Freiberg (GDR) and in Cornwall (UK). (JM^C)

Bournonite *Chemistry* PbCuSbS$_3$. *System* Orthorhombic. *Habit* granular, tabular or prismatic; twinned crystals often look like cog-wheels, hence the name 'wheel-ore'. *Colour* dark grey or black. *Streak* black. *Cleavage* poor. *Fracture* subconchoidal or uneven. *Hardness* 2½–3. *SG* 5.9. *Lustre* metallic. *Special features* high specific gravity and twinned habit. *Formation* with other sulphides in hydrothermal veins. *Distribution* Cornwall (UK), Harz (Germany), Pribram (Czechoslovakia), Huttenberg (Austria), Turin (Italy), Utah (USA), Mexico, Bolivia and Japan. (S)

Tetrahedrite This detailed picture shows the modified tetrahedral crystal faces. (S)

Cobaltite *Chemistry* CoAsS. *System* Cubic. *Habit* cubes, granular, massive. *Colour* silvery steel-grey. *Streak* dark grey. *Cleavage* cubic. *Fracture* uneven. *Hardness* 5½. *SG* 6.3. *Lustre* metallic. *Formation* in hydrothermal veins. *Distribution* Ontario (Canada) and Tunaberg (Sweden) are the main regions. (C)

Galena

Galena

Realgar

Cinnabar

Realgar

Galena

Galena

Galena *Chemistry* PbS. *System* Cubic. *Habit* very often perfect cubes, but can be octahedral. *Colour* lead-grey, though can be brilliant silvery-blue when freshly broken. *Streak* lead-grey. *Cleavage* perfect cubic; when broken many small cubes are produced. *Fracture* even, may be subconchoidal. *Hardness* 2½. *SG* 7.5. *Lustre* metallic to dull; fresh surfaces are brilliant, tarnishes readily to give a dull lustre. *Special features* a very soft metallic mineral with high specific gravity. Perfect cubic crystals are common. When heated with hydrochloric acid, hydrogen sulphide, with its characteristic smell of rotten eggs, is evolved. *Formation* a medium-temperature mineral of hydrothermal veins, galena is found in association with other sulphides such as sphalerite and minerals like quartz, calcite, baryte and fluorite. In some limestones galena occurs as a replacement mineral where it probably originates from disseminated particles leached from veins elsewhere. *Distribution* very common in many areas. The most prolific deposits are in Kansas, Oklahoma and Missouri (USA). It is also found in quantity in Australia at Broken Hill and in Mexico near Chihuahua. Other areas include Italy, Austria, Czechoslovakia, Yugoslavia and Germany. In the UK galena has been mined for hundreds of years in a number of places, notably the northern Pennines, especially Weardale and Alston Moor, Cumbria, Derbyshire, Cornwall and Wanlockhead in southern Scotland. (K, CP)

Cinnabar *Chemistry* HgS. *System* Hexagonal. *Habit* granular and massive, but also tabular and rhombohedral crystals. Can exhibit prismatic and acicular habit. *Colour* deep red to brownish-red. *Streak* scarlet. *Cleavage* perfect prismatic. *Fracture* uneven. *Hardness* 2–2½. *SG* 8.1. *Lustre* adamantine, but dull when massive and at times metallic. *Special features* the colour, streak and very high SG are definitive. *Formation* in association with realgar and pyrite it occurs around hot springs and recent volcanic vents; in joints and faults in sediments. It has been found in deposits of placer type formed from the erosion of ancient mercury-bearing strata. *Distribution* the most famous deposits are near Almanden (Spain) and it is also found at Idria (Yugoslavia), Abbadia San Salvatore (Italy), Nikotowa (USSR), Peru, China; Nevada and Arkansas (USA). (L)

Realgar *Chemistry* AsS. *System* Monoclinic. *Habit* often massive or granular, also as small, short, prismatic crystals. *Colour* deep red to orange. *Streak* red or orange. *Cleavage* good, pinacoidal. *Fracture* uneven or conchoidal. *Hardness* 1½–2. *SG* 3.6. *Lustre* resinous. *Special features* the lustre, hardness and colour are distinctive. *Formation* usually found in the deposits of hot springs and hydrothermal veins, associated with orpiment, tin, antimony, lead and silver. *Distribution* Humbolt County, Nevada, Utah and Washington (USA), Hungary, Yugoslavia, Turkey and Italy. (L, CP)

Jamesonite *Chemistry* $Pb_4FeSb_6S_{14}$. *System* Monoclinic. *Habit* crystals are acicular. *Colour* dark grey. *Streak* greyish-black. *Cleavage* perfect basal. *Fracture* conchoidal and uneven. *Hardness* 2–3. *SG* 5.5–6.0. *Lustre* metallic. *Formation* in hydrothermal veins with lead and iron sulphides. *Distribution* widespread,

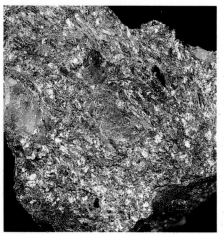

Jamesonite

Niccolite

including Binnental (Switzerland), Yugoslavia, Mexico; Cornwall and the Isle of Man (UK). (C)

Niccolite (Nickeline) *Chemistry* NiAsS. *System* Hexagonal. *Habit* massive, very rarely as short tabular crystals. *Colour* very pale coppery red. *Streak* blackish-brown. *Cleavage* none. *Fracture* uneven. *Hardness* 5–5½. *SG* 7.8. *Lustre* metallic. *Special features* high SG and colour. *Formation* in hydrothermal veins with nickel and silver, also in norite and gabbro with copper and nickel sulphides. *Distribution* Eldorado and Cobalt (Canada), Franklin (USA), Natume (Japan), Jachymov (Czechoslovakia), Harz and Schneeberg (Germany), La Rioja (Argentina) and Italy. (C)

Quartz *Chemistry* SiO$_2$. *System* Trigonal. *Habit* very commonly as six-sided prisms which are terminated by six triangular faces. Many crystals are imperfect and twinning is common. *Colour* one of the most variable minerals, much quartz is white (**milky quartz**) or colourless (**rock crystal**) but the colour varieties are given different names. **Amethyst** which is coloured purple, possibly by manganese or trace amounts of ferric iron, is prized as a semi-precious gemstone. **Rose quartz** is a delicate pink colour due to traces of manganese or titanium, and loses its colour when heated. **Smoky quartz** or **Cairngorm** is a dark brown colour, possibly due to exposure to natural radiation. The very dark or black quartz is called **Morion**. **Citrine** is a variety of quartz with a yellow colour caused by the inclusion of iron hydrates. **Prase** is a green variety, which is so coloured possibly because of the inclusion of minute needles of actinolite. *Streak* white. *Cleavage* none. *Fracture* uneven and conchoidal. *Hardness* 7 (quartz defines point 7 on Mohs scale of hardness). *SG* 2.65. *Lustre* vitreous. *Special features* a very hard mineral with no cleavage and good hexagonal prismatic crystals. *Formation* very common in all manner of geological situations, quartz is a major constituent of many igneous rocks, especially the acid ones such as granite and rhyolite. Many metamorphic rocks like gneiss, schist and meta-quartzite contain large amounts, and some detrital sedimentary rocks are made almost entirely of quartz. It also forms in hydrothermal veins where it is a common gangue mineral. The best crystals are found in cavities such as those in vesicular lavas. *Distribution* worldwide, but some areas have become famous for different varieties. The European Alps of Switzerland and Austria are renowned for rock crystal and smoky quartz; Italy, Czechoslovakia and Romania have provided much amethyst and rock crystal, while Mexico, Brazil and many regions of the USA are a rich source of large amethyst and smoky quartz crystals. Japan is well known for many varieties and Australia has produced some of the best prase. In the UK, Cornwall has amethyst, citrine, prase and smoky quartz, whilst amethyst is also known from Ireland, North Wales, Cumbria and the Mendips. In Scotland, Cairngorm (large crystals come from the mountains of that name), amethyst, rose quartz and much rock crystal are found. (K, BS, JHF, CP)

Quartz from Dauphiné, France A magnificent mass of prismatic crystals. The whole specimen is some 40 cm long. (L)

Cairngorm

Rock crystal

Citrine

Prase

Green quartz

Quartz

Morion

Amethyst

Milky quartz

Rose quartz

Quartz from Dauphiné, France

Agate

Agate

Cristobalite

Agate

Jasper

Chalcedony

Tiger's eye

Agate

Wood opal

Chalcedony

Moss agate

Opal

Chalcedony *Chemistry* SiO_2; this is a micro-crystalline variety of silicon dioxide containing very small crystals and pores. There are a number of named varieties of which chalcedony and **agate** are the main ones. *System* is generally stated as Trigonal, but being micro-crystalline this is not seen with usual observation. *Habit* chalcedony is usually botryoidal or mamillated, agate is banded. *Colour* great variation from white to deep red and blue. The beauty of chalcedony and especially agate is in the juxtaposition of bands of different colours and shades. *Streak* white. *Cleavage* none. *Fracture* conchoidal, uneven. *Hardness* 6½–7. *SG* 2.6. *Lustre* vitreous but many forms of chalcedony have a waxy lustre. *Special features* the habit is characteristic, as is the hardness and higher SG than opal. *Formation* chalcedony and agate form in cavities by precipitation from solutions. Such cavities, sometimes called 'geodes', may contain banded agate with quartz crystals projecting into the open centre. Basaltic lavas commonly contain chalcedony and agate in their vesicular cavities. *Other varieties* **Jasper** is commonly a rich, deep-red colour, due to iron oxide staining. **Moss agate** has a pale background colour with numerous darker dendritic patterns produced by manganese oxide impurities. **Tiger's eye** is a form of quartz with a banded structure and coloured yellow and brown because of fibres of crocidolite. *Distribution* worldwide, though some regions are famous for certain varieties. For example, Brazil, Uraguay and Italy are famous for agate, and some of the best agates in the UK are from the lavas of southern and central Scotland and the Cheviot Hills. South Africa has produced much tiger's eye. (K, BS, JHF, CP)

Cristobalite *Chemistry* SiO_2. *System* Cubic. *Habit* octahedral crystals rare; usually as massive aggregates. *Colour* white or pale grey. *Streak* white. *Cleavage* none. *Fracture* conchoidal. *Hardness* 7. *SG* 2.3. *Lustre* pearly to vitreous. *Special features* hardness and lack of cleavage; the habit helps to distinguish it from quartz. *Formation* as very small crystals in lavas, usually obsidian and other glasses. Cristobalite is the form of silicon dioxide stable above 1470°C. *Distribution* widespread though not very common, it is found in intermediate lavas at Cerro San Cristobal in Mexico, in California, Oregon and Colorado (USA). (K)

Chalcedony 'fingers'

Opal *Chemistry* $SiO_2.nH_2O$; this mineral has no crystal system and occurs in an amorphous form. *Habit* in small veins, rounded globules and stalactitic shapes. *Colour* very variable, from delicate pale blues and pinks to nearly black. Gem quality opal may owe its brilliant play of colours to optical refraction of light among the closely packed spheres of silica in the mineral's structure. *Cleavage* none. *Fracture* conchoidal. *Hardness* 5½–6½. *SG* 1.9–2.5. *Lustre* can be pearly or resinous, but usually vitreous. *Special features* lower SG and hardness than quartz and chalcedony. *Formation* precipitated from silica-rich solutions at low temperatures, in veins and other cavities. It forms at lower temperatures in sedimentary rocks than in volcanic situations. *Distribution* widespread, with certain areas being famous for particular varieties. Australia provides the finest precious opal, in South Australia and New South Wales. Mexico is well known for fire opal, while in the USA, Humboldt County, Nevada produces gem opals, and wood opal is plentiful in Yellowstone National Park, Wyoming. In Europe, Bohemia, Czechoslovakia and Puy-de-Dome (France) are famous localities, and in the UK opals are often found in Devon, Cornwall and Tayside. (L)

Chalcedony 'fingers' Chalcedony often forms in cavities, as here, where a large geode some 20 cm across is filled with 'fingers' of chalcedony. (L)

Magnetite

Zincite

Hematite

Hematite

Chromite

Lamellar magnetite

Cuprite

Hematite

Specular hematite

Specular hematite

Chromite

Magnetite

Franklinite

Spinel

'Iron rose'

Magnetite *Chemistry* Fe$_3$O$_4$; substitutions can take place to produce Franklinite (Zn,Mn,Fe)(Fe,Mn)$_2$O$_4$ Magnesioferrite (MgFe$_2$O$_4$) and Trevorite (NiFe$_2$O$_4$). Magnetite is by far the most common of these. *System* Cubic. *Habit* granular and massive but also forms octahedral and rhombdodecahedral crystals. *Colour* black. *Streak* black. *Cleavage* none, but gives the impression of cleavage by having octahedral partings. *Fracture* uneven to subconchoidal. *Hardness* 5½–6½. *SG* 5.2. *Lustre* metallic to dull. *Special features* magnetite is strongly magnetic, has a dark colour and streak, and is hard. *Formation* in high temperature mineral veins and as an accessory mineral in many basic and ultra-basic rocks. It can occur in large deposits by magmatic segregation. In some metamorphic rocks magnetite is common, and placer sands often yield high amounts of magnetite, the sands being black with the mineral. *Distribution* Kiruna (Sweden) has huge deposits; fine crystals are known from Austria and Switzerland, while considerable deposits are found in the Bushveld complex (South Africa), Utah, Wyoming and the Adirondacks (USA), and the Urals (USSR). (K)

Hematite *Chemistry* Fe$_2$O$_3$. *System* Trigonal. *Habit* tabular or rhombohedral crystals (Specularite when black), also in foliated masses and typically reniform or botryoidal. Can be massive. *Colour* often reddish but can be steely-grey and black. *Streak* red to brownish red. *Cleavage* none. *Fracture* uneven. *Hardness* 5½–6½. *SG* 4.9–5.3. *Lustre* dull to earthy, metallic. *Special features* a hard mineral with a characteristic red streak. The reniform habit is very typical. *Formation* as a hydrothermal mineral and an accessory in igneous rocks. In many sedimentary rocks hematite is of primary occurrence but it also forms large replacement deposits. *Distribution* Lake Superior (USA) is the most productive region; also in Quebec (Canada), Ukraine (USSR) and Cumbria (UK). Many European countries produce good crystals, including Switzerland, Austria, France, Italy and Germany. Brazil and Mexico have fine deposits. (K, CP)

Zincite *Chemistry* ZnO. *System* Hexagonal. *Habit* massive, granular, crystals uncommon. *Colour* red, can be orange-yellow. *Streak* orange-yellow. *Cleavage* prismatic and basal. *Fracture* subconchoidal. *Hardness* 4–4½. *SG* 5.4–5.7.

Lustre subadamantine. *Special features* colour and streak, also soluble in hydrochloric acid. *Formation* in rocks of contact metamorphism. *Distribution* an uncommon mineral, the main occurrence is at Franklin, New Jersey (USA). (K)

Chromite *Chemistry* FeCr$_2$O$_4$. *System* Cubic. *Habit* massive or granular, rarely as octahedral crystals. *Colour* black or brownish. *Streak* brown. *Cleavage* none. *Fracture* uneven. *Hardness* 5½. *SG* 4.5–4.8. *Lustre* submetallic. *Special features* brown streak and weakly magnetic. *Formation* a mineral of ultra-basic rocks such as serpentinite and peridotite where it can be segregated into layers. Chromite also occurs as a placer. *Distribution* much is found in the Urals (USSR), Turkey, Yugoslavia, Austria, Zimbabwe, South Africa, Namibia; Montana, Texas and Pennsylvania (USA). (K)

Cuprite *Chemistry* Cu$_2$O. *System* Cubic. *Habit* octahedral and cubic crystals, rhombdodecahedra and octahedra, massive or granular. *Colour* red to black. *Streak* reddish-brown. *Cleavage* none. *Fracture* uneven, subconchoidal. *Hardness* 3½–4. *SG* 5.8–6.2. *Lustre* adamantine to submetallic. *Special features* quite soft, no cleavage, harder than cinnabar and softer than hematite. *Formation* in the oxidized parts of copper lodes. *Distribution* good crystals are found in Cornwall (UK), near Lyon (France), Arizona and New Mexico (USA), the Urals (USSR), Corocoro (Bolivia) and Namibia. (K)

Franklinite *Chemistry* (Zn,Mn,Fe)(Fe,Mn)$_2$O$_4$. *System* Cubic. *Habit* octahedral crystals, massive or granular. *Colour* black and reddish. *Streak* brownish-red. *Cleavage* none. *Fracture* uneven. *Hardness* 5½–6½. *SG* 5–5.2. *Lustre* metallic. *Formation* in zinc deposits among metasomatic crystalline limestones and dolomites. *Distribution* Franklin and Sterling Hill in Ogdensburg, New Jersey (USA). Here crystals up to 30 cm are found. (L)

Spinel *Chemistry* MgAl$_2$O$_4$. *System* Cubic. *Habit* octahedral, massive, often as rounded grains. *Colour* red, pink, white, green, blue and black. *Streak* white or grey. *Cleavage* none. *Fracture* conchoidal. *Hardness* 8. *SG* 3.5–4.1. *Lustre* vitreous. *Special features* hardness and colour. Colour variations reflect chemical composition, iron, zinc and manganese substituting for magnesium. A series of spinel minerals exists; pleonaste being dark green and containing iron, picotite is brown and contains iron and chromium, gahnite is dark green and contains zinc, hercynite is black and iron-rich. *Formation* in contact metamorphosed limestones and dolomites and in basic to ultra-basic igneous rocks; also in placers. *Distribution* well known in alluvial sands in India, Burma, Thailand, Madagascar, Sri Lanka and Afghanistan. Also found in Italy, Sweden, Canada and the USA. (C)

'Iron rose' This form of hematite has a hexagonal outline and is structured in tabular sheets. (C)

Corundum

Ruby

Cassiterite

Rutile

Siderite

Siderite

Ilmenite

Corundum *Chemistry* Al$_2$O$_3$, *System* Trigonal. *Habit* typically in barrel-shaped prisms, and massive and granular. *Colour* very varied, from colourless to grey and yellow. Blue corundum is **Sapphire** and red is **Ruby**. *Streak* white. *Cleavage* none. *Fracture* conchoidal or uneven. *Hardness* 9. *SG* 3.9–4.1. *Lustre* vitreous to adamantine. *Special features* extreme hardness, habit. *Formation* an accessory mineral in some intermediate rocks such as nepheline syenite, and pegmatites. Also in silica-poor high-grade metamorphic rocks like marble, schist and some gneisses. Because corundum is so hard it also accumulates in alluvial sands. *Distribution* Burma is well known for ruby crystals, while sapphires are found in Sri Lanka, Kashmir, Afghanistan; Queensland and New South Wales (Australia) and Montana (USA). In South Africa large crystals are known from Cape Province. (K)

Cassiterite *Chemistry* SnO$_2$. *System* Tetragonal. *Habit* massive and granular but also in stubby prismatic and pyramidal crystals. *Colour* nearly black or brownish. *Streak* pale grey to white. *Cleavage* prismatic. *Fracture* uneven. *Hardness* 6–7. *SG* 6.8–7.2 *Lustre* sub-metallic, but can be adamantine. *Special features* pale streak and high SG. *Formation* in hydrothermal veins around granite bodies and in alluvial deposits. *Distribution* the USSR, Malaysia and Sumatra (Indonesia) have vast deposits, as do China and Bolivia. Maine and South Dakota (USA), Mexico, Oruro (Bolivia), Germany, France and Portugal all produce fine crystals. Cornwall is the most prolific area in the UK. (CP)

Rutile *Chemistry* TiO$_2$. *System* Tetragonal. *Habit* prismatic with pyramids or massive. *Colour* dark reddish-black or yellowish. *Streak* brown. *Cleavage* prismatic. *Fracture* uneven. *Hardness* 6–6½. *SG* 4.2–4.3. *Lustre* adamantine or metallic. *Special features* habit and lustre. *Formation* as an accessory mineral in a variety of igneous rocks such as granite. In metamorphic gneisses and schists; also as acicular needle-like crystals within quartz. *Distribution* excellent crystals in the Alps of Austria and Switzerland. Also in Norway, the Urals (USSR), Mexico, Brazil and Virginia, Arkansas, North Carolina and Georgia (USA). (K)

Curved faces of siderite Crystal faces are not always flat, plane surfaces. This iron carbonate mineral is seen here exhibiting curved faces on its rhombohedra. Siderite is also known as Chalybite. (L)

Curved faces of siderite

A long-disused tin mine on Craddock Moor, Cornwall, UK

Thin needles of rutile

Cassiterite

Siderite *Chemistry* FeCO$_3$. *System* Trigonal. *Habit* rhombohedra with curved faces, also granular and massive. *Colour* brownish and grey, also black. *Streak* white. *Cleavage* rhombohedral. *Fracture* uneven. *Hardness* 3½–4½. *SG* 3.7–4.0. *Lustre* dull to vitreous. *Special features* curved crystal faces and higher SG than dolomite. Dissolves only very slowly in cold dilute hydrochloric acid. *Formation* in hydrothermal veins, and in clays and shales, often as concretions. Limestones can be replaced by siderite. *Distribution* a widespread and common mineral. Good crystals are found in Hanover (Germany), Styria (Austria), Brosso (Italy), Panasqueriros (Portugal) and Camborne (UK). Greenland, USA, Bolivia, Brazil and Canada have produced much siderite. (K, CP)

Ilmenite *Chemistry* FeTiO$_3$. *System* Trigonal. *Habit* massive or tabular crystals. *Colour* black. *Streak* black to brownish. *Cleavage* none. *Fracture* conchoidal. *Hardness* 5–6. *SG* 4.5–5.0. *Lustre* metallic. *Special features* could be confused with magnetite, but ilmenite is non-magnetic. *Formation* as an accessory mineral in many igneous rocks and also by magmatic segregation. Rarely in quartz veins and pegmatites. As ilmenite is very resistant to weathering it is often found in alluvial sands. *Distribution* France, Switzerland and Norway have provided large crystals. At Travancore in SW India the beach sands contain 70 per cent ilmenite. Also from the Ilmen Mountains (USSR), New York (USA), Canada, Brazil and Cornwall (UK). (K)

Thin needles of rutile A typical occurrence of this titanium-rich mineral is as thin needles embedded in quartz. The term acicular is used to describe this needle-like habit. (L)

Cassiterite The heavy oxide of tin forms very dark coloured crystals. This specimen is in a common association with quartz and is 3 cm across. (C)

A long-disused tin mine on Craddock Moor, Cornwall, UK. This is a monument to the past century, when minerals of tin, lead, zinc and copper were mined at over 600 sites in Cornwall. Specimens of minerals such as sphalerite, galena and cassiterite can be found on the spoil heaps around some of the mines.

Pyrolusite

Brucite

Bauxite

Goethite

Manganite

Psilomelane

Stalactitic and botryoidal goethite

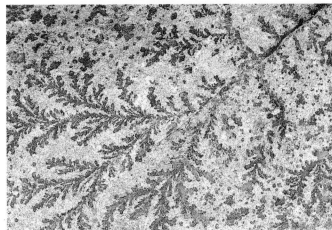

Dendritic pyrolusite

Limonite

Pyrolusite *Chemistry* MnO_2. *System* Tetragonal. *Habit* usually massive, reniform or fibrous, often non-crystalline. *Colour* black. *Streak* black. *Cleavage* prismatic. *Fracture* brittle, uneven. *Hardness* rare crystals 6–6½, massive form 1–2. *SG* 4.7–5.1. *Lustre* metallic or dull. *Special features* very soft when massive, colour. *Formation* often as a precipitate in bogs, lakes and lagoons. Also in nodules on the deep sea bed. *Distribution* great deposits of pyrolusite and wad (manganese oxides) are found in Georgia (USSR), the Deccan (India), Brazil and South Africa. It is mined in USA at Batesville, Arkansas. Czechoslovakia and Germany have produced pyrolusite, as has Cornwall (UK). (K)

Brucite *Chemistry* $Mg(OH)_2$. *System* Trigonal. *Habit* fibrous or tabular crystals, also massive. *Colour* pale green or white. *Streak* white. *Cleavage* perfect basal into sheets or fibres. *Fracture* uneven. *Hardness* 2½. *SG* 2.4. *Lustre* waxy, pearly or vitreous. *Special features* similar to talc but harder, soluble in hydrochloric acid. *Formation* typically found in metamorphosed dolomites and limestones. Also in hydrothermal veins and in chlorite schists and serpentinized rocks. *Distribution* widespread in contact zones, as in Skye and Assynt (UK), also around Vesuvius (Italy); Pennsylvania, Texas and New York (USA), Bajenov (USSR) and Quebec (Canada). (K)

Bauxite *Chemistry* an aggregate of several minerals essentially Al_2O_3. $2H_2O$ with iron oxides. *Habit* massive or concretionary and oolitic masses. *Colour* typically reddish-brown or yellowish. Bauxite does not have distinct properties because of its varied composition. *Formation* by the decay through weathering,

under tropical conditions, of rocks containing silicates of aluminium; heavy tropical rain leaches the silica and aluminium hydroxides remain. *Distribution* Jamaica, Ghana, the USSR and Indonesia all have large deposits of this important ore of aluminium, as do Yugoslavia, France, Italy and Hungary. (K)

Goethite *Chemistry* $FeO(OH)$. *System* Orthorhombic. *Habit* commonly massive or botryoidal but also forms rare prismatic crystals. *Colour* dark brown to black. *Streak* yellowish-brown. *Cleavage* one perfect. *Fracture* uneven. *Hardness* 5–5½. *SG* 3.4–4.3. *Lustre* sub-metallic, fibrous masses silky. *Special features* has a greasy feel; streak, colour and habit characteristic. *Formation* by the oxidation of deposits rich in iron. **Limonite**, which is a yellow-coloured mineral of similar composition to goethite, forms by the weathering of iron deposits. Bog iron ore can contain goethite and limonite. *Distribution* extensive deposits are known in Colorado and Lake Superior region (USA), the USSR, Alsace Lorraine (France), Czechoslovakia, Germany and Cuba. Crystals from Cornwall (UK). (K)

Manganite *Chemistry* $MnO(OH)$. *System* Monoclinic. *Habit* prismatic crystals, often grouped in bundles as radiating masses. *Colour* black or dark grey. *Streak* reddish-brown or black. *Cleavage* perfect pinacoidal, prismatic. *Fracture* uneven. *Hardness* 4. *SG* 4.2–4.4. *Lustre* sub-metallic. *Special features* soluble in concentrated hydrochloric acid, prismatic habit. *Formation* in low temperature hydrothermal veins and in oxidized deposits, often in association with pyrolusite. *Distribution* Ukraine (USSR), Michigan (USA), Harz (GDR) and

many other regions where manganese deposits occur. (K)

Psilomelane *Chemistry* $BaMn^{2+}Mn^{4+}_8O_{16}(OH)_4$. *System* Monoclinic. *Habit* stalactitic, massive, reniform. *Colour* black. *Streak* black. *Cleavage* none. *Fracture* uneven. *Hardness* 5–7. *SG* 3.5–4.7. *Lustre* dull to metallic. *Formation* in the oxidation zone of manganese deposits and in quartz veins. *Distribution* very widespread, large masses at Nikopol (Ukraine) and Chiatura (Georgia, USSR). Also Germany, Austria, South Africa, Ghana, India and Brazil. (L)

Dendritic pyrolusite This oxide of manganese sometimes forms strange plant-like patterns on rock surfaces, which is known as a dendritic habit; this is in great contrast to its occurrence in large masses which are of commercial importance. (C)

Stalactitic and botryoidal goethite Goethite is a dark, quite hard variety of iron hydroxide which occurs as massive specimens and commonly in the stalactitic or botryoidal habit, both of which are seen in this specimen. (C)

Limonite has a similar composition to goethite but is yellow in colour and is amorphous. Its main constituent is microcrystalline goethite but it may also contain much ochreous yellow ferric oxide. It is widespread and occurs as a secondary mineral in weathering zones but large masses of limonite and goethite develop through precipitation in both marine and fresh water and in bogs. (S)

Fibrous magnesite

Magnesite

Dolomite

Dolomite

Rhodochrosite

Smithsonite

Dolomite rhombs

Rhodochrosite showing a reniform habit

Magnesite *Chemistry* MgCO₃. *System* Trigonal. *Habit* massive, granular or fibrous; rhombohedral or prismatic crystals rare. *Colour* white, but when impurities like iron are present it can be yellowish and brown. *Streak* white. *Cleavage* rhombohedral. *Fracture* conchoidal. *Hardness* 3½–4½. *SG* 3.0. *Lustre* vitreous. *Special features* only reacts with dilute hydrochloric acid when warm, otherwise rather like calcite, but slightly harder and denser. *Formation* as a replacement mineral and in talc schists and serpentinites, where carbonated water has reacted with serpentinite which is rich in magnesium. *Distribution* extensive deposits are in the Urals (USSR), California and Washington (USA), Austria, Manchuria (China), Poland, India and Greece. (K)

Dolomite *Chemistry* CaMg(CO₃)₂. *System* Trigonal. *Habit* rhombohedral crystals with curved faces; also massive and granular. *Colour* white, but often brownish. *Streak* white. *Cleavage* rhombohedral. *Fracture* conchoidal, uneven. *Hardness* 3½–4. *SG* 2.8–2.9. *Lustre* pearly or vitreous. *Special features* curved crystal faces and very slow reaction with cold dilute hydrochloric acid distinguish dolomite from calcite. *Formation* as a hydrothermal mineral and also by the replacement of limestones by the action of magnesium-rich solutions. *Distribution* a very common mineral, fine crystals are known from Salzburg (Austria), Binnental (Switzerland), Freiberg (GDR), Brosso (Italy), Cornwall and Cumbria (UK), Trepca (Yugoslavia); Missouri, Iowa, New Jersey and Michigan (USA), Brazil and Mexico. (K, JHF)

Smithsonite *Chemistry* ZnCO₃. *System* Trigonal. *Habit* typically reniform or botryoidal, very rarely as rhombohedral crystals. Also as encrusting and stalactitic masses. *Colour* grey, white and brownish but also bright green and blue. *Streak* white. *Cleavage* rhombohedral. *Fracture* uneven. *Hardness* 5½. *SG* 4.3–4.5. *Lustre* vitreous. *Special features* carbon dioxide is liberated with effervescence when cold hydrochloric acid is added; high SG, reniform habit. *Formation* in the oxidized zone of zinc deposits, in some hydrothermal veins and as a replacement in limestones. *Distribution* in many zinc and lead mining regions. Fine specimens have been found in Colorado, Arizona and Arkansas (USA), the USSR, Bolivia, Brazil, Turkey, Erzberg (Austria), Italy, France and Tsumeb (Namibia). In the UK smithsonite is known from the Mendips, Matlock, Leadhills and Alston Moor. (K)

Rhodochrosite *Chemistry* MnCO₃. *System* Trigonal. *Habit* typically massive or granular but rhombohedral crystals occur with curved faces. *Colour* light brown or grey but characteristically pale pink to deep reddish-pink. *Streak* white. *Cleavage* rhombohedral. *Fracture* uneven. *Hardness* 3½–4½. *SG* 3.3–3.7. *Lustre* vitreous. *Special features* colour, effervesces in cold dilute hydrochloric acid. *Formation* in hydrothermal veins often with copper, lead and silver sulphides. Also in some metamorphosed sediments. *Distribution* very fine crystals have been found at Colorado (USA), Romania, Freiberg (GDR), Hotazel (South Africa), Chiatura (USSR), Argentina, Mexico and Italy. (CP)

Dolomite rhombs Dolomite often occurs with other hydrothermal or replacement minerals. This specimen shows cream dolomite rhombs, with curved faces, with two forms of hematite; brownish-red reniform and black specularite.

Malachite

Azurite

Malachite

Pseudomalachite

Malachite

Cerrusite

Malachite

Witherite

Malachite *Chemistry* $Cu_2CO_3(OH)_2$. *System* Monoclinic. *Habit* typically as banded botryoidal masses, rarely as acicular crystals. *Colour* green. *Streak* light green. *Cleavage* good, pinacoidal. *Fracture* uneven or subconchoidal. *Hardness* 3½–4. *SG* 4.0. *Lustre* crystals are adamantine, broken surfaces of botryoidal masses are silky. *Special features* effervesces with dilute hydrochloric acid, green colour, botryoidal habit. *Formation* usually in the oxidation zone of copper deposits, also disseminated in sandstones by meteoric waters. *Distribution* a common copper mineral, large amounts have been found in Shaba (Zaire), Zambia, Otavi (SW Africa), Zimbabwe, the Urals and Siberia (USSR). Also in New South Wales and South Australia, Utah and Arizona (USA), Chessy (France), Romania, Germany, Italy, Chile, Cornwall and Cumbria (UK). (K, CP)

Pseudomalachite *Chemistry* $6CuO.P_2O_5.3H_2O$. *System* Monoclinic. *Habit* crystals rare, massive or in aggregates. *Colour* dark emerald green. *Streak* dark green. *Cleavage* good in one direction. *Fracture* splintery, uneven. *Hardness* 4½–5. *SG* 4.4. *Lustre* vitreous. *Special features* colour, soluble in hydrochloric acid. *Formation* associated with malachite in the oxidized zones of copper deposits. *Distribution* similar to malachite, well known from Collier Bay (Western Australia), Cornwall (UK) and Pennsylvania (USA). (L)

Cerrusite *Chemistry* $PbCO_3$. *System* Orthorhombic. *Habit* usually prismatic crystals, which typically form a radiating network. Also massive and granular. *Colour* grey or white. *Streak* white. *Cleavage* prismatic. *Fracture* conchoidal, brittle. *Hardness* 3–3½. *SG* 6.5. *Lustre* adamantine. *Special features* high SG, lustre, dissolves in nitric acid whereas anglesite ($PbSO_4$) does not. *Formation* a secondary mineral formed in the oxidized zones of lead deposits. *Distribution* a widespread mineral, fine crystals from New Mexico, Colorado and Pennsylvania (USA), Tsumeb (Namibia), Friedrichssegen and Mechernich (Germany), Mezica (Yugoslavia), Czechoslovakia, Broken Hill (Australia), Kazakhstan (USSR), Italy and Sardinia; Cornwall, Derbyshire, Leadhills and Durham (UK). (K)

Witherite *Chemistry* $BaCO_3$. *System* Orthorhombic. *Habit* pseudo-hexagonal crystals, usually twinned. Also massive or botryoidal. *Colour* white, grey or yellowish. *Streak* white. *Cleavage* pinacoidal. *Fracture* uneven. *Hardness* 3½. *SG* 4.3. *Lustre* vitreous or resinous. *Special features* effervesces in cold dilute hydrochloric acid, high SG. *Formation* in hydrothermal veins associated with baryte and galena. *Distribution* Illinois and California (USA), Siberia (USSR), Sardinia; Alston Moor, Hexham and North Wales (UK). (K)

Azurite *Chemistry* $Cu_3(CO_3)_2(OH)_2$. *System* Monoclinic. *Habit* often massive but crystals may be prismatic and grouped in radiating aggregates. *Colour* dark blue. *Streak* pale blue. *Cleavage* prismatic. *Fracture* conchoidal, a brittle mineral. *Hardness* 3½–4. *SG* 3.8. *Lustre* vitreous. *Special features* the dark blue colour is distinctive. Azurite effervesces in nitric and hydrochloric acids. *Formation* a secondary mineral which occurs in the oxidized zones of copper deposits, often with malachite. *Distribution* the finest crystals are from Chessy (France), hence the alternative name chessylite. Also from Tsumeb (Namibia), Laurium (Greece), Broken Hill (Australia), Chile,

Malachite and azurite

Aurichalcite

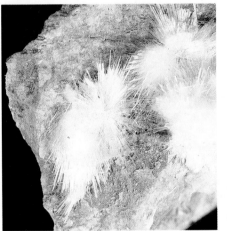

Artinite

Prismatic crystals of cerrusite

Siberia, Iran, Mexico and Arizona (USA). In the UK from Redruth, Caldbeck fells and Galway. (K, C)

Artinite *Chemistry* $Mg_2(CO_3)(OH)_2.3H_2O$. *System* Monoclinic. *Habit* acicular and rounded aggregates. *Colour* grey or white. *Streak* white. *Cleavage* perfect. *Fracture* uneven. *Hardness* 2. *SG* 2.0. *Lustre* silky. *Special features* habit, hardness, soluble in weak HCl with effervescence. *Formation* in hydrothermal veins and serpentinized ultra-basic rocks. *Distribution* New Jersey and New York (USA), Italy, Yugoslavia and Austria. (C)

Aurichalcite *Chemistry* $(Zn,Cu)_5(CO_3)_2(OH)_6$. *System* Monoclinic. *Habit* in crusts and mamillated nodules; acicular

crystals. *Colour* pale blue-green. *Streak* white. *Cleavage* good, breaks into flexible plates. *Fracture* uneven. *Hardness* 2–2½. *SG* 4.2. *Lustre* pearly to vitreous. *Special features* a very soft mineral with distinctive colour and moderately high SG. *Formation* with sulphides of copper and zinc, commonly in the oxidized zones of hydrothermal veins. *Distribution* Leadhills (Scotland), Chessy (France), Laurium (Greece), Italy, Tsumeb (Namibia), Utah and New Mexico (USA). (CP)

Malachite and azurite often occur together as in this specimen from the Atlas mountains of Morocco. Both these minerals have a similar chemistry but the green colour of malachite distinguishes it from the rich blue of azurite. (C)

Halite

Halite

Sylvine

Halite

Halite

Sylvine

Halite

Cryolite

Carnallite

Halite (Rock Salt) *Chemistry* NaCl. *System* Cubic. *Habit* cubes, very often with hollow faces which are hopper crystals. Also granular and massive. *Colour* colourless, white, yellow and red, grey and blue. *Streak* white. *Cleavage* cubic. *Fracture* uneven. *Hardness* 2½. *SG* 2.2. *Lustre* vitreous. *Special features* distinct salty taste, soluble in water, cubic cleavage. *Formation* an evaporite mineral formed when bodies of saline water dry out. *Distribution* large deposits in Utah, Texas, Louisiana and New York (USA), Stassfurt (Germany), Cardona (Spain), Wieliczka and Bochnia (Poland), Salzkammergut (Austria), Iletskaya Zashchita (Siberia, USSR) and China. In the UK much halite is found in Cheshire, North Yorkshire, Cleveland and Durham. (K)

Sylvine (Sylvite) *Chemistry* KCl. *System* Cubic. *Habit* cubes and octahedra often combined. Also massive. *Colour* white, grey or pale reddish. *Streak* white. *Cleavage* cubic. *Fracture* uneven. *Hardness* 2. *SG* 2.0. *Lustre* vitreous. *Special features* tastes more bitter than halite. *Formation* as an evaporite, associated with halite, but less frequent than halite because it is more soluble. *Distribution* similar to that of halite. Stassfurt (Germany), Texas, Utah and New Mexico (USA), Ukraine (USSR), Kalusz (Poland), Cardona (Spain) and Saskatchewan (Canada). In UK in Cleveland. (K)

Cryolite *Chemistry* Na_3AlF_6. *System* Monoclinic. *Habit* massive or granular, pseudo-cubic crystals very rare. *Colour* white, brown or reddish, can be black. *Streak* white. *Cleavage* none. *Fracture* uneven. *Hardness* 2½. *SG* 3.0. *Lustre* vitreous or greasy. *Special features* becomes invisible when placed in water because it has the same refractive index. *Formation* in pegmatite dykes in granite, with galena, quartz and fluorite. *Distribution* an unusual mineral, the main occurrence of which is at Ivigtut in Greenland. Also recorded from Colorado (USA), Montreal (Canada), Nigeria, and Miass (USSR). (K)

Carnallite *Chemistry* $KMgCl_3.6H_2O$. *System* Orthorhombic. *Habit* massive and granular, crystals very rare. *Colour* white or reddish-brown. *Streak* white. *Cleavage* none. *Fracture* conchoidal. *Hardness* 2. *SG* 1.6. *Lustre* greasy. *Special features* deliquescent and deteriorates on exposure to air by taking in water. Bitter taste. *Formation* an evaporite mineral, occurring with halite, sylvite and other evaporites. It is one of the last precipitates from saline solutions. *Distribution* very similar to the other evaporites, Texas and New Mexico (USA), Stassfurt (Germany), Iran, China, Solikamsk (USSR) and Cleveland (UK). (CP)

Fluorite (Fluorspar) *Chemistry* CaF_2. *System* Cubic. *Habit* cubic crystals are very common, also as rhombdodecahedra and octahedra; twinned crystals with interlocking cubes. Fluorite also occurs in banded and concretionary forms. *Colour* a great variety of colours are found, ranging from purple and blue to green, yellow, reddish and white. Transparent colourless crystals are not

Blue John

Specimens of various forms of fluorite

uncommon and black fluorite is occasionally found. *Streak* white. *Cleavage* perfect cleavage planes parallel to the octahedron are produced, having the effect of making triangular planes across the corners of cube-shaped crystals. *Fracture* uneven and conchoidal. *Hardness* 4 (this mineral is point 4 on Mohs scale). *SG* 3.3. *Lustre* vitreous. *Special features* the cube-shaped crystals and octahedral cleavage are distinctive, especially in combination with the typical green or purple colour. The bluish-purple banded variety is called Blue John and is much used as an ornamental stone, being carved into vases and other decorative shapes. *Formation* in hydrothermal mineral veins in association with galena, sphalerite, baryte, quartz, calcite and

others. It also occurs in very small amounts in some granites and rarely as a cement in detrital sediments. *Distribution* an extremely widespread mineral. Very fine crystals are known from many areas, including Illinois, Kentucky, Ohio and New Hampshire (USA), Ontario (Canada), Mexico; Harz, Wolfach and Freiberg (Germany), Grimsel and Fellital (Switzerland), Bolzano (Italy), Kongsberg (Norway), Cera (Brazil) and the USSR. In the UK many areas produce fine fluorite, including Weardale, Wensleydale, Alston Moor, Castleton (Derbyshire), which is the main source of Blue John, the Mendip Hills, Denbighshire, Cumbria, Wanlockhead in south Scotland and Cornwall. (K, CP, JM^C)

Stalactite

Aragonite

Calcite

Calcite

Calcite

Iceland spar

Stalagmite

Nail-head calcite

Calcite

Atacamite *Chemistry* $Cu_2Cl(OH)_3$. *System*
Orthorhombic. *Habit* thin prismatic crystals,
acicular, radiating and concretionary. *Colour*
green. *Streak* green. *Cleavage* pinacoidal.
Fracture conchoidal. *Hardness* 3–3½. *SG* 3.8.
Lustre vitreous. *Special features* colour and
habit. *Formation* in the weathered and oxidized
parts of copper lodes. *Distribution* Atacama
(Chile), Peru, Bolivia, Mexico, Wallaroo
(Australia), Namibia; Arizona and Utah (USA),
Kazakhstan and the Urals (USSR), Italy, and
Cornwall (UK). (L)

Calcite *Chemistry* $CaCO_3$. *System* Trigonal.
Habit a great variety of crystal forms is known.
Prismatic, tabular, scalenohedral and
rhombohedral habits are common; 'nail-head'
crystals are combinations of prism and
rhombohedron, 'dog-tooth' crystals are prisms
combined with scalenohedra. Calcite shows
greater variations in its habit than any other
mineral, and twinning is also common. Massive,
stalactitic, fibrous and granular habits also
occur. *Colour* very varied. White and colourless
calcite is common, but green, yellow, grey,
purple, red and brown forms are known. *Streak*
white. *Cleavage* perfect, rhombohedral.
Fracture very difficult to see because of the
perfect cleavage, but conchoidal. *Hardness* 3
(calcite is the standard of 3 on Mohs scale). *SG*
2.7. *Lustre* vitreous to pearly. *Special features*
effervesces in cold dilute hydrochloric acid,
liberating carbon dioxide. The perfect cleavage
is very characteristic. Clear cleavage rhombs
produce double refraction. Originally known
from Iceland (and therefore called Iceland spar),
such clear rhombs produce two images of a
single object when placed over it. The specimen
figured has a single line beneath it, which
appears double when seen through the rhomb.
Formation in many geological situations calcite
is a common mineral. Hydrothermal veins often
contain fine calcite crystals in association with
baryte, fluorite and sulphides. Limestone and
its metamorphic equivalent, marble, contain a
very high proportion of calcite, and where lime-
rich waters run through underground cave
systems, calcite is deposited as stalactites and
stalagmites. Calcite forms around hot springs
and in evaporites. The igneous rocks called
carbonatites are of magmatic origin and contain
much calcite. *Distribution* a very widespread
mineral, fine crystals are known from many
regions including: Harz and St Andreasberg
(Germany), Bleiberg (Austria), Switzerland,
Kapnik (Hungary), Rhisnes (Belgium),
Fontainebleau (France), Helgustadir and
Eskifjord (Iceland); Connecticut, New Jersey,
Keeweenaw (Michigan), Kansas, Missouri,
Oklahoma, New York, Arizona, Indiana and
Idaho (USA); Guanajuato, Naica and
Chihuahua (Mexico), Tsumeb and
Grootgontein (SW Africa). In the UK many
regions provide excellent calcite crystals.
Notable among these are Bontallack, Levant
and Geevor mines in Cornwall; the Derbyshire
mining area; the Mendips and South Wales;
Egremont and Frizington in West Cumbria; the
northern Pennine ore field; Leadhills and
Wanlockhead in southern Scotland. (K, CP)

Aragonite *Chemistry* $CaCO_3$. *System*
Orthorhombic. *Habit* pseudo-hexagonal
twinned crystals are more common than
untwinned prismatic ones, which are often
acicular. Also occurs in coralloid, stalactitic and
fibrous habits. *Colour* white, grey, pinkish, pale
yellow, green, blue, red and colourless. *Streak*
white. *Cleavage* poor, pinacoidal. *Fracture*
subconchoidal. *Hardness* 3½–4. *SG* 2.9. *Lustre*
vitreous. *Special features* effervesces strongly

with cold dilute hydrochloric acid, harder and
denser than calcite, no rhombic cleavage.
Formation far less common than calcite,
aragonite occurs in deposits from hot springs
and in evaporites. In metamorphic rocks it
occurs in some glaucophane schists.
Occasionally in the oxidized zones of ore
deposits. *Distribution* fine specimens are known
from Molina de Aragon (Spain), Bastennes
(France), Bilina (Czechoslovakia), Dognacska
(Hungary); Salzberg, Erzberg and Werfen
(Austria), Tarnowitz (Poland), Agrigento
(Sicily); Arizona and New Mexico (USA). In the
UK, fine crystals are known from the Alston
Moor area. (K)

Strontianite *Chemistry* $SrCO_3$. *System*
Orthorhombic. *Habit* acicular crystals in

bundles, also massive and granular. *Colour*
white, pale green, grey, yellow or pink. *Streak*
white. *Cleavage* prismatic. *Fracture* uneven.
Hardness 3½–4. *SG* 3.7. *Lustre* vitreous. *Special
features* soluble in dilute hydrochloric acid.
Formation in hydrothermal veins. *Distribution*
Strontian, Alston Moor, Hexham and Weardale
(UK); California and Illinois (USA), Westphalia
(Germany), Salzburg (Austria) and Siberia
(USSR). (C)

Aragonite Two specimens showing the
pseudo-hexagonal habit, from Aragon,
Spain. (C)

Calcite A specimen showing the
scalenohedron habit. (C)

Atacamite

Strontianite

Calcite

Aragonite

Baryte

Colemanite

Baryte

Baryte

Baryte

Baryte

Baryte

Celestine

Borax

Barytocalcite

Borax *Chemistry* $Na_2B_4O_7.10H_2O$. *System* Monoclinic. *Habit* short prismatic crystals, massive. *Colour* white or yellowish. *Streak* white. *Cleavage* perfect pinacoidal. *Fracture* conchoidal. *Hardness* 2–2½. *SG* 1.7. *Lustre* greasy or dull. *Special features* soluble in water. *Formation* an evaporite. *Distribution* Death Valley, Searles Lake, Clear Lake and Boron (California, USA), Tibet, Lop Nor and Kashmir (central Asia), especially on lakeshores. (C)

Barytocalcite *Chemistry* $BaCO_3.CaCO_3$. *System* Monoclinic. *Habit* prismatic. *Colour* white to yellowish-brown. *Streak* white. *Cleavage* prismatic. *Fracture* uneven. *Hardness* 4. *SG* 3.6. *Lustre* resinous or vitreous. *Formation* in hydrothermal veins. *Distribution* well known from lead veins in northern England. (C)

Celestine (Celestite) *Chemistry* $SrSO_4$. *System* Orthorhombic. *Habit* prismatic and tabular crystals; can be massive and granular. *Colour* colourless and white, or yellowish, reddish and pale blue. *Streak* white. *Cleavage* prismatic and basal. *Fracture* uneven. *Hardness* 3–3½. *SG* 3.9. *Lustre* vitreous. *Special features* similar to baryte but has lower SG. *Formation* in hydrothermal veins and in cavities in volcanic rocks; in evaporite deposits with anhydrite and in dolomitic limestones. *Distribution* Stadtberge (Germany), Biel (Switzerland), Agrigento (Sicily), Perticara (Italy), Bleiberg (Austria); Lake Erie, Ohio, Utah and Lampasas, Texas (USA). Yate near Bristol is a famous locality in the UK. (K)

Baryte (Barite, Barytes) *Chemistry* $BaSO_4$. *System* Orthorhombic. *Habit* usually tabular, but prismatic crystals also occur, as do lamellar, fibrous, cockscomb and 'desert rose' forms. *Colour* colourless, white, brown, reddish, blue and green. *Streak* white. *Cleavage* basal and prismatic. *Fracture* uneven. *Hardness* 2½–3½. *SG* 4.5. *Lustre* vitreous. *Special features* high SG, habit. *Formation* commonly found in hydrothermal veins with a variety of other minerals including sulphides, fluorite, calcite, quartz and dolomite. In limestones as a replacement and in sandstones as cement. *Distribution* a very common mineral in many regions: Bleiberg, Huttenberg and Schwarz (Austria); Baumholder, Freiburg and Munsteral (Germany); Binnental and Valais (Switzerland),

Ulexite

Gypsum

Polyhalite

Selenite

Satin spar

Selenite

Desert rose

Daisy gypsum

Anhydrite

Bologna (Italy), Pribram (Czechoslovakia); Dakota, California, Oklahoma and Colorado (USA); Cumbria, Durham, Cornwall and Derbyshire (UK). (K)

Colemanite *Chemistry* $Ca_2B_6O_{11}.5H_2O$. *System* Monoclinic. *Habit* short prismatic crystals, massive and granular. *Colour* white, pink, red, grey and yellowish. *Streak* white. *Cleavage* one perfect. *Fracture* uneven. *Hardness* 4–4½. *SG* 2.4. *Lustre* vitreous. *Formation* in cavities in sediments and in evaporite basins, probably from waters which have passed through borate deposits. *Distribution* Death Valley in California (USA), Chile, Argentina, Kazakhstan (USSR) and Bigadic (Turkey). (K)

Ulexite *Chemistry* $NaCaB_5O_9.8H_2O$. *System* Triclinic. *Habit* spongy, rounded masses or fibrous. *Colour* white. *Streak* white. *Cleavage* microscopic prismatic. *Fracture* uneven. *Hardness* 1–2½. *SG* 2.0. *Lustre* silky. *Special features* habit, hardness, soluble in hot water. *Formation* in evaporite deposits with borax and colemanite. *Distribution* Argentina, Peru, Atacama (Chile); Boron, Mojave desert (USA) and Italy. (C)

Gypsum *Chemistry* $CaSO_4.2H_2O$. *System* Monoclinic. *Habit* tabular, crystals often have swallowtail twinning. Clear transparent crystals are called **selenite**, while rosette-shaped masses, often with much sand in them, are called **desert roses**. **Satin spar** is a fibrous form; massive and granular habit occur, the latter being **alabaster**. **Daisy gypsum** consists of fine radiating crystals. *Colour* white and colourless, also pink, brown, grey and yellow. *Streak* white. *Cleavage* three cleavage planes; that parallel to the pinacoid produces very thin flexible but inelastic plates. *Fracture* uneven. *Hardness* 2 (gypsum is the standard on the Mohs scale). *SG* 2.3. *Lustre* vitreous, pearly on cleavage faces. *Special features* very soft, can be scratched with a fingernail. *Formation* an evaporite mineral, associated with halite and carnallite. Also occurs as a precipitate from hot springs and in beds of clay as selenite crystals. *Distribution* a common mineral. Large accumulations are found in many regions in evaporite sequences, for example in north-east England, Germany and USA. Fine crystals are known from Montmartre (France), Agrigento (Sicily); Arzberg, Oberfranken and Demmelswald (Germany), Chihuahua (Mexico) and Utah (USA). Desert

roses are common in New Mexico, Arizona (USA), Tunisia and Morocco. The USSR, USA, Canada, UK, Spain, Italy and Germany produce much gypsum commercially. In the UK, fine selenite crystals are found in many clay beds such as the Oxford and London clays. (K, CP)

Anhydrite *Chemistry* $CaSO_4$. *System* Orthorhombic. *Habit* massive or granular, crystals very rare. *Colour* white or colourless, may be bluish or grey. *Streak* white. *Cleavage* three at right-angles. *Fracture* uneven. *Hardness* 3–3½. *SG* 2.9. *Lustre* pearly or vitreous. *Special features* cleavage and hardness distinguish

anhydrite from gypsum. *Formation* as an evaporite mineral and by the dehydration of gypsum. Often in salt domes as a cap-rock and very rarely in hydrothermal veins. *Distribution* as for other evaporites. (CP)

Polyhalite *Chemistry* $K_2Ca_2Mg(SO_4)_4.2H_2O$. *System* Triclinic. *Habit* lamellar or fibrous. *Colour* reddish and pink. *Streak* white. *Clea*[vage] parallel to the pinacoid. *Fracture* uneve[n.] *Hardness* 2½–3. *SG* 2.7. *Lustre* re[sinous,] silky. *Special features* has a b[rick-?] red appearance. *Formatio[n]* *Distribution* as for other eva[porites.]

Linarite

Glauberite

Crocoite

Dioptase

Almandine

Glauberite *Chemistry* Na_2SO_4. $CaSO_4$. *System* Monoclinic. *Habit* tabular, prismatic, dipyramidal. *Colour* greyish, yellowish, red-brown. *Streak* white. *Cleavage* perfect basal. *Fracture* conchoidal. *Hardness* 2½–3. *SG* 2.8. *Lustre* vitreous to greasy. *Formation* an evaporite. *Distribution* Italy and Sicily, Lorraine (France), Ebro (Spain), Salzburg (Austria), Ruthenia (USSR), New Mexico, California, Arizona and Texas (USA). (C)

Linarite *Chemistry* $(Pb,Cu)_2SO_4(OH)_2$. *System* Monoclinic. *Habit* prismatic, acicular or tabular crystals. *Colour* deep azure blue. *Streak* pale blue. *Cleavage* perfect pinacoidal. *Fracture* uneven. *Hardness* 2½. *SG* 5.4. *Lustre* vitreous. *Formation* in the alteration zone of copper and lead deposits. *Distribution* Arizona, Utah and California (USA), Linares (Spain), Sardinia, Namibia; Cumbria and Leadhills (Scotland). (C)

Crocoite *Chemistry* $PbCrO_4$. *System* Monoclinic. *Habit* prismatic crystals, often in acicular masses; massive or granular. *Colour* reddish-orange to brown. *Streak* yellowish. *Cleavage* prismatic. *Fracture* uneven, brittle. *Hardness* 2½–3. *SG* 5.9–6.0. *Lustre* vitreous. *Special features* colour and high SG. *Formation* a secondary mineral formed in the oxidized parts of lead veins. *Distribution* Nontron (France), Pennsylvania (USA), Dundas (Tasmania), Goyabeira (Brazil), the Urals (USSR) and Labo (Philippines). (C)

Almandine This garnet is common in micaschists and in this specimen from Norway the dark almandine crystals are set in schist rich in muscovite mica. (C)

Olivine *Chemistry* $(Mg,Fe)_2SiO_4$. The term olivine refers to a continuous series of minerals from Forsterite, Mg_2SiO_4, to Fayalite, Fe_2SiO_4. The properties of olivines vary as the iron content increases. Gem-quality forsterite is called peridot. *System* Orthorhombic. *Habit* rarely as crystals, commonly as grains and granular masses. *Colour* pale olive-green to yellow and brown. *Streak* colourless. *Cleavage* very poor pinacoidal. *Fracture* conchoidal. *Hardness* 6½–7. *SG* 3.2 (forsterite)–4.3 (fayalite). *Lustre* vitreous. *Special features* a hard mineral with distinctive colour and habit. *Formation* olivine is a main constituent of basic and ultra-basic igneous rocks such as basalt, gabbro and peridotite. Dunite is an ultra-basic rock composed almost entirely of olivine. Forsterite is the olivine with this occurrence and is also found in metamorphosed dolomites. Fayalite is far less common and forms in some acid igneous rocks which have cooled very rapidly, for example pitchstone and rhyolite. Olivine is also found in lunar basalt, and nickel-rich forsterites occur in some meteorites. *Distribution* a widespread mineral; basalt is an abundant rock, covering the ocean floors. Gem-quality olivine is known from Burma, Sri Lanka, Zebirget (Egypt), Norway, Eifel (Germany), Arizona (USA), Mt Vesuvius (Italy) and the Urals (USSR). (K, CP)

Garnet *Chemistry* the general formula for this group of minerals is $A_3B_2Si_3O_{12}$. A can be Ca, Mn, Mg or Fe, and B is usually Al, but can be Cr or Fe. A number of more common garnets are named and can be distinguished by colour and other features. **Pyrope** $Mg_3Al_2Si_3O_{12}$, blood-red in colour, often very dark. **Almandine** $Fe_3Al_2Si_3O_{12}$, dark brownish-red, purplish-black. **Spessartine** $Mn_3Al_2Si_3O_{12}$, yellowish-orange. **Grossular** $Ca_3Al_2Si_3O_{12}$, colourless, greenish, yellow, red. Hessonite is an orange form. **Andradite** $Ca_3Fe_2Si_3O_{12}$, green, brown and black. The black form is melanite, the green is demantoid. **Uvarovite** $Ca_3Cr_2Si_3O_{12}$, an uncommon green form. *System* Cubic. *Habit* well-formed crystals very common. Rhombdodecahedra and trapezohedra and more complex modifications. Also massive and granular. *Colour* see above. *Streak* greyish. *Cleavage* none. *Fracture* uneven to subconchoidal. *Hardness* 6½–7½. *SG* 3.5–4.3, according to composition. *Lustre* vitreous or resinous. *Special features* a very hard mineral with well-formed crystals common. Colour is a fair method of distinguishing different garnets and specific gravity will help further. Only a detailed chemical test will give exact identification of varieties. *Formation* common in many metamorphic and some igneous rocks. These include schists, gneisses and marbles, pegmatites, serpentinites and peridotites. In sediments garnet occurs in river and beach sands from the breakdown of garnet-bearing primary rocks. *Distribution* a widespread and common mineral; the following areas are noted for fine crystals: Zermatt, Binnental, Valais, Graubunden and Ticino (Switzerland), Carinthia, Tirol, Granatkogel and Zillertal (Austria), Auerbach, Aschaffenburg and Wurlitz (Germany), Bilina (Czechoslovakia), Nizhni-Tagil (USSR); Alto Adige, Traversella and Val Malenco (Italy), Outokumpu (Finland), Quebec (Canada), Chihuahua (Mexico), many parts of the USA, including New Jersey and New York, Idaho, Alaska, Pennsylvania and California; also Tanzania and Kenya (Africa), Australia and Greenland. In the UK garnets are common in the schists and gneisses of the Scottish Highlands. (K, CP, C)

Dioptase *Chemistry* $CuSi_2(OH)_2$. *System* Trigonal. *Habit* stubby, prismatic crystals, also massive. *Colour* bright emerald green. *Streak* green. *Cleavage* perfect rhombohedral. *Fracture* uneven or conchoidal. *Hardness* 5. *SG* 3.3. *Lustre* vitreous. *Special features* colour, hardness. *Formation* in the oxidized parts of copper veins. *Distribution* well-known localities for crystals are Tsumeb (Namibia), Mindonli (Zaire), Copiapo and Atacama (Chile), Arizona (USA) and Kazakhstan (USSR). (C)

Andradite The greenish variety of this type of garnet is called demantoid, seen here as a mass of twinned crystals. (C)

Olivine

Garnet

Olivine

Olivine

Garnet

Garnet

Garnet

Andradite

Kyanite

Kyanite

Tourmaline

Tourmaline

Tourmaline in granite

Topaz

Andalusite variety chiastolite

Kyanite *Chemistry* Al₂SiO₅. *System* Triclinic. *Habit* thin, blade-like crystals and radiating masses. *Colour* white, light blue, grey or greenish. Often the crystals are darker colours in the centre. *Streak* white. *Cleavage* good in two directions. *Fracture* uneven. *Hardness* 6–7 across the cleavage planes, 4–5 along them. *SG* 3.6. *Lustre* vitreous or pearly. *Special features* habit, colour and variation in hardness. *Formation* mainly in medium- to high-grade regionally metamorphosed schists and gneisses. Also in pegmatites in these rocks. *Distribution* of wide occurrence, fine crystals from: Laufenberg and Greiner (Austria), Passo Cristallina and Pizzo Forno (Switzerland), Morbihan (France), Musso and Passiria (Italy), Sultan Hamud (Tanzania), Machakos (Kenya), Minas Gerais (Brazil), Sri Lanka, India, Australia and Massachusetts, Connecticut and North Carolina (USA). In the UK kyanite is not uncommon in the Scottish Highlands and fine specimens are known from Mull, Orkney and Kinloch Rannoch. (K, CP)

Tourmaline *Chemistry* Na(Mg,Fe,Li,Mn, Al)₃Al₆(BO₃)₃Si₆O₁₈(OH,F)₄. *System* Trigonal. *Habit* prismatic, occasionally massive. *Colour* black or very dark blue. Also pink, green and colourless. Often crystals are pink at one end and green at the other. *Streak* colourless. *Cleavage* very poor, rhombohedral. *Fracture* uneven or conchoidal. *Hardness* 7. *SG* 2.9–3.2. *Lustre* vitreous. *Special features* colour, habit with striations lengthwise on the prisms. *Formation* mainly in granites, pegmatites and greisens. Also in some gneisses and schists, and as detrital grains in sediments. *Distribution* fine tourmalines are known from: Spittal (Austria), Dobrava and Drava River (Yugoslavia), Kragero (Norway), Elba (Italy), Mursinka (the Urals, USSR), Madagascar, Lourenco (Mozambique), Nepal, Sri Lanka, Minas Gerais (Brazil), Mexico, Yinnietharra (W Australia) and many parts of the USA including California, Maine, New York and New Jersey. In the UK tourmaline is common in the granites of Devon and Cornwall and also occurs in Cumbria (Carrock mine) and many of the Scottish granites, for example in the Cairngorms and near New Galloway in the Southern Uplands. (K, CP)

Topaz *Chemistry* Al₂SiO₄ (OH,F)₂. *System* Orthorhombic. *Habit* prismatic crystals, granular or massive. *Colour* yellowish, greenish, blue or reddish-brown. *Streak* white. *Cleavage* perfect basal. *Fracture* conchoidal or uneven. *Hardness* 8. *SG* 3.6. *Lustre* vitreous. *Formation* in pegmatites in greisen, granites and rhyolites. *Distribution* very large crystals (up to 270 kg) from Minas Gerais (Brazil). Also from Saxony (Germany), Elba (Italy), Omi (Japan), Klein

Spits Kopje (SW Africa), California, Utah and Colorado (USA), Ukraine and the Urals (USSR) and Cornwall (UK). (L)

Hemimorphite *Chemistry* Zn₄Si₂O₇(OH)₂.H₂O. *System* Orthorhombic. *Habit* tabular crystals, also often fibrous or botryoidal. *Colour* white, blue, brown or yellow. *Streak* white. *Cleavage* prismatic. *Fracture* uneven or conchoidal. *Hardness* 4½–5. *SG* 3.5. *Lustre* vitreous. *Formation* in the oxidation zone of zinc veins. *Distribution* common in many regions. Fine material from Saxony (Germany), Sardinia (Italy), Belgium, Austria, Algeria, Madagascar, Chihuahua (Mexico); New Jersey, Missouri and Pennsylvania (USA), Kazakhstan (USSR), Cumbria and Derbyshire (UK). (L)

Staurolite *Chemistry* Fe₂Al₉Si₄O₂₂(OH)₂. *System* Monoclinic, may have higher symmetry of Orthorhombic system. *Habit* short prismatic crystals or massive. *Colour* brown to black. *Streak* greyish. *Cleavage* one good. *Fracture* uneven. *Hardness* 7–7½. *SG* 3.7. *Lustre* resinous or vitreous. *Special features* often forms cross-shaped twins. *Formation* in medium-grade metamorphic rocks, also in placers. *Distribution* widespread, fine crystals from Otzal Alps (Austria), Ticino (Switzerland), Kleinosthein (Germany), Como (Italy), Sanarka river (USSR) and Scotland. (L)

Idocrase (Vesuvianite) *Chemistry* Ca₁₀Mg₂Al₄(SiO₄)₅(Si₂O₇)₂(OH)₄. *System* Tetragonal. *Habit* prismatic crystals, massive and granular. *Colour* dark greenish or brown. *Streak* white. *Cleavage* very poor. *Fracture* conchoidal, brittle. *Hardness* 6–7. *SG* 3.3–3.4. *Lustre* resinous or vitreous. *Formation* in impure limestones which have suffered contact metamorphism. *Distribution* widespread, Monte Somma and Vesuvius (hence other name) (Italy), Zermatt (Switzerland), the Urals, Arenal (Norway), California (USA), Loch Tay (Scotland). (C)

Andalusite and Chiastolite *Chemistry* Al₂SiO₅. *System* Orthorhombic. *Habit* short prismatic crystals with square cross-section, also massive. Chiastolite has dark carbonaceous inclusions arranged to form a cross shape. *Colour* red, pink, brown, green or grey. *Streak* white. *Cleavage* prismatic. *Fracture* uneven. *Hardness* 6½–7½. *SG* 3.2. *Lustre* vitreous. *Formation* in low pressure metamorphic rocks. *Distribution* widespread, large crystals from Andalusia (Spain), also from Minas Gerais (Brazil), California and Pennsylvania (USA), Bimbowrie (Australia), Senus-Bugu (USSR) and many parts of the UK including Cumbria and Scotland. (C)

Hemimorphite

Staurolite

Idocrase

Cordierite

Nepheline

Beryl

Clinozoisite

Stilbite

Sodalite

Lazurite

Emerald

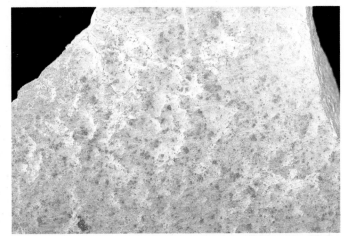

Zoisite variety Thulite

Epidote

Beryl *Chemistry* $Be_3Al_2Si_6O_{18}$. *System* Hexagonal. *Habit* prismatic crystals, sometimes of great size, up to 9m weighing over 25.4 tonnes, have been recorded. Also occurs massive. *Colour* commonly pale greyish or greenish but transparent gem-quality beryl exhibits bright green (**Emerald**), yellow (**Heliodor**), blue (**Aquamarine**) and pink (**Morganite**). *Streak* white. *Cleavage* very indistinct basal. *Fracture* uneven or conchoidal. *Hardness* 7½–8. *SG* 2.6–2.9. *Lustre* vitreous to resinous. *Special features* a very hard, prismatic mineral. *Formation* in granites and pegmatites, also in some mica schists and gneisses. Alluvial sediments derived from these rocks will contain beryl because of its hardness and chemical stability. *Distribution* Austria, Switzerland, Italy and Siberia, Gravelotte (S Africa), Zimbabwe, Madagascar, India, many parts of the USA, Takovaya River (USSR), Japan and Australia. (K)

Cordierite *Chemistry* $(Mg,Fe)_2Al_4Si_5O_{18}$. *System* Orthorhombic. *Habit* short pseudo-hexagonal crystals, also massive and granular. *Colour* very dark blue. *Streak* colourless. *Cleavage* one, poorly displayed. *Fracture* uneven or subconchoidal. *Hardness* 7–7½. *SG* 2.6–2.8. *Lustre* vitreous. *Special features* great hardness, colour and habit. *Formation* in medium- to high-grade metamorphic rocks such as schists, gneisses and contact hornfelses. *Distribution* widespread, but good crystalline material rare. Fine specimens known from Bavaria, Finland, Norway, Sweden, India, Brazil, Canada and the USA. (K)

Clinozoisite (Monoclinic Epidote) *Chemistry* $Ca_2Al_3Si_3O_{12}(OH)$. *System* Monoclinic. *Habit* prismatic, granular, massive or fibrous. *Colour* grey, greenish, pinkish. *Streak* white. *Cleavage* one, lengthwise. *Fracture* uneven. *Hardness* 6–7. *SG* 3.2–3.4. *Lustre* vitreous. *Formation* in regionally metamorphosed low- to medium-grade rocks, also in some contact metamorphosed limestones. *Distribution* Bern, Guttanen (Switzerland), Hohe Tauern and Untersulzbachtal (Austria), Val di Fassa and Elba (Italy), Dauphiné (France), Arendal (Norway), SE Alaska; Colorado, California, Connecticut (USA), Los Gavilanes and Baja California (Mexico) and Mozambique. (K)

Nepheline *Chemistry* $NaAlSiO_4$. *System* Hexagonal. *Habit* prisms and grains. *Colour* white, grey, reddish, greenish, brown. *Streak* white. *Cleavage* basal. *Fracture* conchoidal. *Hardness* 5½–6. *SG* 2.6. *Lustre* vitreous or greasy. *Special features* greasy lustre. *Formation*

in intermediate igneous rocks such as syenite and phonolite. *Distribution* Kola pèninsula (USSR), Mt Vesuvius (Italy) and South Africa, all noted for large masses; a widespread mineral. (K)

Stilbite *Chemistry* $NaCa_2(Al_5Si_{13})O_{36}.14H_2O$. *System* Monoclinic. *Habit* in radiating or sheaf-like aggregates. *Colour* white to reddish or yellowish. *Streak* white. *Cleavage* one, perfect. *Fracture* uneven. *Hardness* 3½–4. *SG* 2.1. *Lustre* vitreous, cleavage surfaces are pearly. *Special features* habit and lustre. *Formation* in cavities and hollows in basaltic lavas. *Distribution* fine specimens from Elba (Italy), Rio Grande do Sul (Brazil), Teigarhorn (Iceland), New Jersey (USA), Nova Scotia (Canada), Poona (India) and Skye (Scotland). (K)

Sodalite *Chemistry* $Na_8Al_6Si_6O_{24}Cl_2$. *System* Cubic. *Habit* usually massive, rare crystals rhombdodecahedral. *Colour* bright azure blue, also yellow, greenish and white. *Streak* colourless. *Cleavage* rhombdodecahedral. *Fracture* uneven or conchoidal. *Hardness* 5½–6. *SG* 2.2. *Lustre* vitreous. *Special features* distinctive colour. *Formation* in igneous rocks of alkaline type, such as nepheline syenite; also in silica-deficient lavas. *Distribution* Mt Vesuvius, Naples and Viterbo (Italy), Burma, Bancroft (Canada), Arkansas (USA), Bolivia, Brazil, the USSR, Greenland, Romania, Portugal and Zimbabwe. (JHF)

Lazurite *Chemistry* $(Na,Ca)_8(Al,Si)_{12}O_{24}(S,SO_4)$. *System* Cubic. *Habit* rare cubic crystals, commonly massive. *Colour* rich azure blue. *Streak* bluish-white.

Cleavage rhombdodecahedral. *Fracture* uneven. *Hardness* 5–5½. *SG* 2.4. *Lustre* vitreous. *Special features* colour, often speckled with iron pyrites. *Formation* in marbles formed by contact metamorphism with granites. Lapis lazuli is the name given to rocks rich in lazurite. *Distribution* Lake Baikal (USSR), Kokscha valley (Afghanistan), Burma, Italy, Pakistan, Labrador (Canada), Colorado and California (USA). (K)

Emerald The bright green variety of beryl which is of gem quality is seen here as two crystals, about 2 cm long, set in a mass of quartz. This specimen is from Bogota, Colombia. (L)

Epidote The specimen shows a small prismatic crystal of this monoclinic mineral. Epidote is a group name for a number of minerals including clinozoisite and zoisite. (L)

Zoisite (variety Thulite) *Chemistry* $Ca_2Al_3Si_3O_{12}(OH)$. *System* Orthorhombic. *Habit* prismatic crystals are elongated; massive. *Colour* the variety of zoisite illustrated is thulite, which is pink in colour because of manganese in its structure. Other forms are brown and grey. *Cleavage* perfect in one direction. *Fracture* uneven. *Hardness* 6. *SG* 3.3. *Lustre* pearly where cleaved, otherwise vitreous. *Formation* in medium- to high-grade metamorphic rocks such as schist, gneiss and eclogite. *Distribution* thulite is not uncommon in manganese-bearing rocks, as at Telemark (Norway), Sondrio and Bolzano (Italy) and South Carolina (USA). Zoisite in its other forms is widespread. (C)

Hornblende

Tremolite

Hornblende

Cummingtonite

Hornblende

Glaucophane

Actinolite

Hornblende *Chemistry*
$(Ca,Na)_{2-3}(Mg,Fe,Al)_5(Si,Al)_8O_{22}(OH)_2$.
System Monoclinic. *Habit* prismatic, granular
and fibrous. *Colour* green or black. *Streak*
brownish. *Cleavage* prismatic, two intersecting
at 60° and 120°. *Fracture* uneven. *Hardness* 5–6.
SG 3–3.4. *Lustre* vitreous. *Special features*
colour and cleavage intersection. *Formation*
common in igneous rocks, especially
hornblende granites, granodiorites, syenites and
some gabbros. Also in hornblende schists and
amphibolites. *Distribution* widespread and
common. (K, CP)

Tremolite-Actinolite *Chemistry* a series
ranging from $Ca_2Mg_5Si_8O_{22}(OH)_2$ (Tremolite),
to $Ca_2(Mg,Fe)_5Si_8O_{22}(OH)_2$ (Actinolite). *System*
Monoclinic. *Habit* thin prisms, radiating or
fibrous. *Colour* grey to green, occasionally
pinkish. *Streak* white. *Cleavage* perfect
prismatic. *Fracture* uneven. *Hardness* 5–6. *SG*
2.9–3.4, depending on iron content. *Lustre*
vitreous. *Special features* fibrous habit.
Formation tremolite is characteristic of contact
metamorphosed dolomites and in serpentinites.
Actinolite forms in schists and amphibolites,
often from the metamorphism of basic igneous
rocks. *Distribution* a common series, fine
specimens from Val Germanasca (Italy),
Zillertal and Obergurgl (Austria), Campolongo
and Poshiaro (Switzerland), Lake Baikal
(USSR), Jade Mt (Alaska), Wyoming,
California (USA), Canada, New Zealand, Kuen
Lun Mountains (China) and west Sutherland
(Scotland). (K, CP,L)

Cummingtonite *Chemistry*
$(Fe,Mg)_7Si_8O_{22}(OH)_2$. *System* Monoclinic.
Habit fibrous aggregates. *Colour* grey, green and
brownish. *Cleavage* prismatic. *Fracture* uneven.
Hardness 5–6. *SG* 3.2–3.7. *Lustre* vitreous to
silky. *Formation* in regionally metamorphosed
schists and in some contact metamorphosed
rocks. *Distribution* widespread; Cummington,
Massachusetts (USA) is a well-known
locality. (K)

Glaucophane *Chemistry*
$Na_2(Mg,Fe,Al)_5Si_8O_{22}(OH)_2$. *System*
Monoclinic. *Habit* fibrous and acicular. *Colour*
grey, blue-black. *Cleavage* prismatic. *Fracture*
uneven. *Hardness* 6–6½. *SG* 3–3.3. *Lustre*
vitreous. *Formation* in low temperature, high
pressure metamorphic rocks. *Distribution* fine
material from: Langesundfjord and Svalbard
(Norway), Falun (Sweden), Azerbaidzhan
(USSR), Prieska (S Africa), Valais (Switzerland)
and California (USA). (K)

Pectolite *Chemistry* $Ca_2NaHSi_3O_9$. *System*
Triclinic. *Habit* acicular crystals, often in
radiating aggregates. *Colour* white or grey.
Streak white. *Cleavage* perfect in two directions.
Fracture uneven. *Hardness* 4½–5. *SG* 2.9. *Lustre*
silky or vitreous. *Special features* habit and
hardness. *Formation* with zeolites in vesicular
cavities in lavas, especially basalts. *Distribution*
widespread, good specimens from New Jersey
(USA), Quebec and Ontario (Canada), Kola
peninsula (USSR), Antrim (Ireland),
Czechoslovakia, and Trento (Italy). (C)

Neptunite *Chemistry*
$Na_2KLi(Fe,Mn)_2Ti_2Si_8O_{24}$. *System* Monoclinic.
Habit prismatic. *Colour* black. *Cleavage* perfect.
Fracture uneven. *Hardness* 5–6. *SG* 3.2. *Lustre*
vitreous. *Special features* colour and habit.
Formation in undersaturated plutonic rocks
such as nepheline syenites, as an accessory
mineral. *Distribution* California (USA), Kola
peninsula (USSR), Ireland and Greenland. (C)

Pectolite

Tremolite

Neptunite

MINERALS

Hypersthene

Wollastonite

Spodumene

Spodumene

Kunzite

Augite

Hedenbergite

Rhodonite

Petalite

Diopside

Monoclinic. *Habit* stubby prismatic crystals. *Colour* greenish-black. *Streak* greyish-green. *Cleavage* prismatic, two planes intersecting at 86°. *Fracture* uneven. *Hardness* 5–6½. *SG* 3.2–3.6. *Lustre* vitreous. *Special features* cleavage, habit and colour. *Formation* common in basic and ultra-basic igneous rocks such as gabbro, basalt and pyroxenite. *Distribution* very widespread, fine specimens from many igneous masses including Bushveldt (S Africa), Stillwater (USA) and Skærgaard (Greenland). (CP)

Diopside (a clinopyroxene) *Chemistry* CaMg,FeSi$_2$O$_6$. *System* Monoclinic. *Habit* prismatic crystals, often in granular or fibrous aggregates. *Colour* paler green than augite, also brown or yellowish. *Streak* greyish. *Cleavage* prismatic, intersection as for augite. *Fracture* uneven. *Hardness* 5–6. *SG* 3.4. *Lustre* vitreous. *Formation* in impure limestones which have suffered contact metamorphism, also in basalts. *Distribution* very widespread, fine crystals from Trento and Turin (Italy), Binnental (Switzerland), Tirol and Zillertal (Austria), Finland, the Urals (USSR), Sala (Sweden), Greenland, New Zealand, New York (USA) and Scotland. (C)

Hedenbergite (a clinopyroxene) *Chemistry* CaFeSi$_2$O$_6$. *System* Monoclinic. *Habit* radiating aggregates are common. *Colour* dark grey to black. *Streak* brownish. *Cleavage* prismatic. *Fracture* uneven. *Hardness* 5–6. *SG* 3.6. *Lustre* vitreous. *Formation* in contact metamorphosed limestones and iron-rich sediments. *Distribution* radiating aggregates from Cornwall (UK), Kazakhstan and Altai (USSR), Australia, Nigeria, Obira (Japan), Arendal (Norway), Elba and Livorno (Italy). (K)

Spodumene (a clinopyroxene) *Chemistry* LiAlSi$_2$O$_6$. *System* Monoclinic. *Habit* prismatic. *Colour* white, grey, yellowish, pink (**kunzite**), green (**hiddenite**). *Streak* white. *Cleavage* prismatic. *Fracture* uneven. *Hardness* 6½–7. *SG* 3.2. *Lustre* vitreous. *Formation* in granitic pegmatites. *Distribution* very large crystals up to 13 m from South Dakota (USA), Manitoba (Canada) and the Urals (USSR). Fine kunzite from San Diego, California (USA) and Brazil. Hiddenite from N Carolina (USA) and Brazil, Madagascar, Mexico, Sweden, Peterhead (Scotland) and Leinster (southern Ireland). (K)

Wollastonite *Chemistry* CaSiO$_3$. *System* Triclinic. *Habit* tabular and prismatic; fibrous and radiating. *Colour* grey or white. *Streak* white. *Cleavage* in three directions. *Fracture* uneven. *Hardness* 4½–5. *SG* 2.8–3.1. *Lustre* vitreous. *Special features* cleavage, soluble in HCl. *Formation* in metamorphosed limestones, both contact and high-grade regional. *Distribution* widespread, fine specimens from Vesuvius and Capo di Bove (Italy), Brittany (France), the Black Forest (Germany), Csiklowa (Romania), California and New York (USA) and Chapas (Mexico). (K)

Rhodonite *Chemistry* (Mn,Fe,Ca)SiO$_3$. *System* Triclinic. *Habit* commonly massive or granular; rare as tabular crystals. *Colour* pink and brown. *Streak* white. *Cleavage* prismatic. *Fracture* uneven, a brittle mineral. *Hardness* 5½–6½. *SG* 3.4–3.7. *Lustre* vitreous. *Special features* similar to rhodochrosite but harder. *Formation* in hydrothermal veins and metamorphosed manganese-rich limestones. *Distribution* Broken Hill (Australia), Arrow Valley (New Zealand), Pajsberg and Langban (Sweden), Sverdlovsk

Hypersthene and Enstatite (orthopyroxenes) *Chemistry* (Mg,Fe)SiO$_3$ and MgSiO$_3$. *System* Orthorhombic. *Habit* rarely as prismatic crystals, often massive or granular. *Colour* greenish-brown or black. *Streak* white. *Cleavage* prismatic. *Fracture* uneven. *Hardness* 5–6. *SG* 3.1–4.0, depending on iron content. *Lustre* vitreous. *Formation* found in basic and ultra-basic igneous rocks, some intermediate volcanic rocks and meteorites. *Distribution* widespread, large masses at Lake Baikal (USSR) and the Cortland complex (USA). (K)

Augite (a clinopyroxene) *Chemistry* (Ca,Mg,Fe,Ti,Al)(Al,Si)$_2$O$_6$. *System*

(USSR), New Jersey (USA), Genoa and Sondrio (Italy), Japan, South Africa and Brazil. (K)

Petalite *Chemistry* $LiAlSi_4O_{10}$. *System* Monoclinic. *Habit* rare tabular crystals, cleaved masses common. *Colour* greenish, reddish or grey. *Streak* white. *Cleavage* one, perfect. *Fracture* conchoidal. *Hardness* 6–6½. *SG* 2.4. *Lustre* pearly on cleavages, vitreous. *Formation* in granitic pegmatites with tourmaline, spodumene and lepidolite. *Distribution* well-known localites include Norwich, Massachusetts and Maine (USA), Varutrask (Sweden), Bickita (Zimbabwe); also in S Africa, Australia and Peru. (K)

Muscovite *Chemistry* $KAl_2(AlSi_3O_{10})(OH,F)_2$. *System* Monoclinic. *Habit* tabular with pseudohexagonal outline; foliated scaly masses. *Colour* silvery, colourless or white, transparent. *Streak* white. *Cleavage* basal into elastic, flexible flakes. *Fracture* uneven. *Hardness* 2½–3. *SG* 2.8. *Lustre* vitreous. *Special features* habit, cleavage and colour are distinctive. *Formation* in acid igneous rocks, often in large masses in pegmatites and granites. Also in schists and gneisses and some detrital sediments. *Distribution* a common, widespread mineral; in Ontario (Canada) crystals measuring 30–50 sq m occur in pegmatites. In Nellore (India) a crystal 3 × 5 m weighed 86.3 tonnes. (K)

Biotite *Chemistry* $K(Mg,Fe)_3AlSi_3O_{10}(OH,F)_2$.
Phlogopite *Chemistry* $KMg_3(AlSi_3)O_{10}(OH,F)_2$.
System Monoclinic. *Habit* tabular with pseudohexagonal outline, small flakes and platy aggregates. *Colour* biotite is black or dark brown, phlogopite is brownish-yellow. *Streak* greyish-white. *Cleavage* basal into flakes. *Fracture* uneven. *Hardness* 2–3. *SG* 2.8–3.2. *Lustre* vitreous, cleavage surfaces are metallic when weathered. *Special features* habit, colour and cleavage. *Formation* common in many igneous rocks, especially acid ones; also in schists and gneisses. *Distribution* a common mineral, very large crystals are recorded from the Ilmen mountains (USSR), Brazil and Greenland. (K, CP)

Lepidolite *Chemistry* $K(Li,Al)_3(Si,Al)_4O_{10}(OH,F)_2$. *System* Monoclinic. *Habit* flakes or platy masses. *Colour* grey, pinkish or lilac. *Cleavage* basal. *Fracture* uneven. *Hardness* 2½–4. *SG* 2.8–2.9. *Lustre* vitreous or pearly. *Special features* colour distinguishes lepidolite from other micas. *Formation* usually in granite rocks with spodumene and tourmaline. *Distribution* very large masses at Minas Gerais (Brazil) and Pala (California, USA); also in Connecticut, New Mexico and South Dakota (USA), Madagascar, Zimbabwe, the Urals (USSR), Coolgardie (Australia), Uto (Sweden) and Japan. (K)

Chlorite *Chemistry* $(Mg,Fe,Al)_6(Si,Al)_4(OH)_8$. *System* Monoclinic. *Habit* flaky, pseudohexagonal crystals, massive. *Colour* greenish, brown and yellow. *Streak* pale green. *Cleavage* basal into non-elastic flakes. *Fracture* uneven. *Hardness* 1½–2½. *SG* 2.5–3.2. *Lustre* vitreous. *Special features* colour and non-elastic cleavage flakes. *Formation* many silicate minerals alter to chlorite. Also found in low-grade metamorphic rocks including slates and phyllites. *Distribution* common and widespread; very large crystals from Ontario (Canada) and Putnam County (New York, USA). (K)

Chlorite

Muscovite

Lepidolite

Phlogopite

Biotite

Biotite

Kaolinite *Chemistry* $Al_2Si_2O_5(OH)_4$. *System* Monoclinic or Triclinic. *Habit* massive, microscopic hexagonal. *Colour* white, pink or grey. *Streak* white. *Cleavage* perfect basal. *Fracture* uneven. *Hardness* 2–2½ (often less when massive). *SG* 2.5. *Lustre* dull when massive, pearly in minute flakes. *Formation* an alteration product of feldspars, both through weathering and on a larger scale by hydrothermal activity. *Distribution* an extremely common and widespread mineral; very large masses in Saxony (Germany), Limoges (France), Czechoslovakia, the Ukraine and the Urals (USSR), Virginia and Georgia (USA), Janchu Fa (China) and Cornwall (UK). (C)

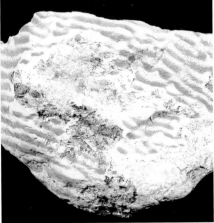

Kaolinite

Serpentine

Prehnite

Serpentine

Eudialyte

Talc

Chrysotile

Prehnite

Talc

Sphene

Pyrophyllite

Antigorite

Apophyllite

Serpentine *Chemistry* $Mg_3Si_2O_5(OH)_4$. *System* Monoclinic. *Habit* lamellar (**Antigorite**), fibrous (**Chrysotile**). *Colour* brown, green, grey, yellow, white or red. *Cleavage* chrysotile has none, antigorite is basal. *Fracture* conchoidal. *Hardness* 2½–4. *SG* 2.5. *Lustre* pearly or greasy, silky. *Special features* colour and lustre. *Formation* as an alteration product of olivine, pyroxenes and amphiboles. *Distribution* widespread and common, fine decorative material from Cornwall (UK), Kemmleten (Switzerland), Bernstein (Austria); New Jersey, California and Arizona (USA) and Quebec (Canada). (K)

Prehnite *Chemistry* $Ca_2Al_2Si_3O_{10}(OH)_2$. *System* Orthorhombic. *Habit* usually reniform or fibrous. *Colour* pale green. *Streak* white. *Cleavage* basal. *Fracture* uneven. *Hardness* 6–6½. *SG* 2.9. *Lustre* vitreous. *Special features* colour and reniform habit. *Formation* in fissures in igneous rocks and in some low-grade metamorphosed rocks. *Distribution* Rhine–Palatinate (Germany), Dauphiné (France); Keeweenaw, Michigan, New Jersey, Lake Superior and Virginia (USA), Renfrew and Dumbarton (Scotland). (K)

Talc *Chemistry* $Mg_3Si_4O_{10}(OH)_2$. *System* Monoclinic. *Habit* massive or tabular. *Colour* pale greenish-grey, white. *Streak* white. *Cleavage* basal. *Fracture* uneven. *Hardness* 1 (talc defines this point on the Mohs scale). *SG* 2.8. *Lustre* pearly or dull. *Special features* very soft and has a greasy or soapy feel (soapstone). *Formation* an alteration product of ferro-magnesian minerals in ultra-basic rocks, also found in schists and marbles. *Distribution* common, but well-known localities include Styria (Austria), Goptersgrun (Germany), Bamle and Snarum (Norway), Italy, Madras (India), Australia, Japan, Georgia, North Carolina and California (USA) and Canada. (K, CP)

Sphene *Chemistry* $CaTiSiO_5$. *System* Monoclinic. *Habit* massive or short wedge-shaped crystals. *Colour* brown, also grey and greenish. *Streak* white. *Cleavage* prismatic. *Fracture* conchoidal. *Hardness* 5–5½. *SG* 3.5. *Lustre* adamantine or resinous. *Special features* habit, colour and lustre are distinctive. *Formation* in igneous rocks such as syenite and diorite as an accessory, also in metamorphosed limestones and schists. *Distribution* large deposits on the Kola peninsula (USSR), crystals at Salzburg (Austria), Turin (Italy), Binnental (Switzerland), Ontario (Canada), New York and New Jersey (USA), San Quentin (Mexico), South Africa and Japan. (K)

Eudialyte *Chemistry* $Na_4(Ca,Fe)_2ZrSi_6O_{17}(OH,Cl)_2$. *System* Trigonal. *Habit* tabular, rhombohedral, massive. *Colour* pinkish-red. *Streak* white. *Cleavage* poor, basal. *Fracture* uneven. *Hardness* 5½. *SG* 3.0. *Lustre* vitreous. *Special features* colour quite distinctive. *Formation* as an accessory in certain intermediate igneous rocks like syenites. *Distribution* well known from the Kola peninsula (USSR), South Africa, Japan and the USA. (LW)

Pyrophyllite *Chemistry* $Al_2Si_4O_{10}(OH)_2$. *System* Monoclinic. *Habit* radiating or lamellar aggregates. *Colour* green, pale yellow or brownish. *Streak* white. *Cleavage* perfect into flexible fragments. *Fracture* uneven. *Hardness* 1–2. *SG* 2.8. *Lustre* pearly. *Special features* very soft, greasy to touch. *Formation* as masses in crystalline schists, often with talc, from which it can be difficult to distinguish. *Distribution* Scotland, Switzerland, Austria, Sicily, S Africa; North and South Carolina, Arkansas and Georgia (USA) and the southern Urals (USSR). (C)

Apophyllite *Chemistry* $KFCa_4Si_8O_{20}.8H_2O$. *System* Tetragonal. *Habit* a variety of crystals including prisms, octahedra, pseudo-cubes and dipyramids. *Colour* white, yellow, grey, green or pink. *Streak* white. *Cleavage* poor prismatic and good basal. *Fracture* uneven. *Hardness* 4½–5. *SG* 2.4. *Lustre* vitreous but pearly on cleavage surfaces. *Special features* pearly lustre on cleavage planes, basal faces can be rough. *Formation* hydrothermal veins and with zeolites in vesicular lavas. *Distribution* Scotland, Iceland, Faeroes, Greenland, Bolzano (Italy), Andreasberg (Germany), Nova Scotia (Canada), Lake Superior and New Jersey (USA), Brazil and Poona (India). (C)

Antigorite is a lamellar variety of serpentine. (L)

Zircon

Zircon *Chemistry* $ZrSiO_4$. *System* Tetragonal. *Habit* short prismatic crystals, often pyramidal at each end; also granular. *Colour* brown, red, grey, green or yellow. *Streak* white. *Cleavage* poor, prismatic. *Fracture* conchoidal. *Hardness* 7½. *SG* 4.7. *Lustre* vitreous. *Special features* very hard, usually reddish-brown. *Formation* a common accessory mineral of many igneous rocks, also in schist and gneiss. Rounded grains occur in placers. *Distribution* widespread, large crystals from Maine (USA), Renfrew (Canada), Norway, Sweden, Australia and Miass (USSR). Alluvial zircon is common in Florida (USA), Brazil, Matura (Sri Lanka) and the Urals (USSR). (C)

Plagioclase *Chemistry* $NaAlSi_3O_8$ – $CaAl_2Si_2O_8$. Different forms of plagioclase occur with the changing ratio of Na and Ca in the plagioclase chemistry. **Albite** is the name given to the Na-rich variety, and at the other end of the series is Ca-rich **anorthite**. **Oligoclase** has about 80 per cent Na and 20 per cent Ca; **andesine** 60 per cent Na and 40 per cent Ca; **labradorite** 40 per cent Na and 60 per cent Ca and **bytownite** 20 per cent Na and 80 per cent Ca. It is very difficult to tell these types of plagioclase without microscopic thin-section examination. *System* Triclinic. *Habit* prismatic or tabular, also massive. *Colour* white, grey, greenish, rarely pink. Labradorite can be told by its bright blue, purple and green colours. *Streak* white. *Cleavage* two planes, easily distinguished. *Fracture* uneven. *Hardness* 6–6½. *SG* 2.6–2.75, increasing with Ca content. *Lustre* vitreous. *Special features* a characteristic feature of plagioclase is the repeated twinning which is best seen on cleavage surfaces. *Formation* an essential and very common mineral in igneous rocks, plagioclase is one of the parameters of classification of these rocks. Sodium-rich plagioclase is common in the acid rocks, such as granite, while in the basic rocks, including gabbro and basalt, calcium plagioclase occurs. Metamorphic rocks contain plagioclase, as do rocks recovered from the Moon's surface, and detrital sediments. *Distribution* one of the most common minerals, abundant in many regions. Very large crystals are known from Labrador (Canada) and Japan. (K, JHF, CP)

Orthoclase *Chemistry* $KAlSi_3O_8$. *System* Monoclinic. *Habit* prismatic or tabular. *Colour* white, grey, pink, green (**amazonstone**). *Streak* white. *Cleavage* two, very good. *Fracture* uneven or conchoidal. *Hardness* 6. *SG* 2.6. *Lustre* vitreous, pearly on cleavage surfaces. *Special features* simple twinning of orthoclase helps to distinguish it from plagioclase. Orthoclase is often pink, the green colour of amazonstone is characteristic. *Varieties* **Microcline** is a Triclinic form of orthoclase which under the microscope shows characteristic 'cross-hatched' patterns because of repeated twinning. **Adularia** is a low temperature form. **Anorthoclase** has a chemical structure of $(Na,K)AlSi_3O_8$. *Formation* common in many igneous rocks, especially granites and acid pegmatites. Also in schists and gneisses and in immature sandstones as detrital grains. *Distribution* a very common widespread mineral. Microcline crystals 30 m long are known from Bavaria, and yellow orthoclase from Madagascar is used as a gem-stone. Blue crystals are known from Lake Baikal (USSR) and large pale crystals have been found at Tokyo Bay (Japan). (K, BS)

Albite

Labradorite

Albite

Plagioclase

Amazonstone

Amazonstone

Microcline

Orthoclase

Orthoclase

Orthoclase

Adularia

Anorthoclase

Analcime

Chabazite

Natrolite

Harmotome

Zeolites are silicate minerals which contain water of crystallization. They are frequently found in cavities in basalts and other volcanic rocks and in hydrothermal situations, including veins. Five examples are illustrated.

Analcime *Chemistry* $NaAlSi_2O_6.H_2O$. *System* Cubic. *Habit* trapezohedral crystals; also massive and granular. *Colour* white. *Streak* white. *Cleavage* cubic. *Fracture* subconchoidal. *Hardness* 5½. *SG* 2.3. *Lustre* vitreous. *Formation* associated with other zeolites in cavities in lavas and intrusive igneous rocks, as a secondary mineral. *Distribution* widespread; fine crystals from Sicily, Italy, Germany; Colorado, Michigan and New Jersey (USA), Canada, Iceland, Tasmania, Sydney (Australia) and Skye (Scotland). (C)

Chabazite *Chemistry* $(Ca,Na_2)Al_2Si_4O_{12}.6H_2O$. *System* Trigonal. *Habit* rhombohedral crystals with pseudocubic appearance. *Colour* white, red or greenish. *Streak* white. *Cleavage* rhombic. *Fracture* uneven. *Hardness* 4½. *SG* 2.0. *Lustre* vitreous. *Formation* with other zeolites in hollows in basalts. *Distribution* fine crystals from Sardinia, Italy, Nova Scotia (Canada), New Jersey (USA), Scotland, Ireland, Faeroes, Germany and Czechoslovakia. (C)

Natrolite *Chemistry* $Na_2Al_2Si_3O_{10}.2H_2O$. *System* Orthorhombic. *Habit* aggregates of acicular radiating crystals. *Colour* white. *Streak* white. *Cleavage* prismatic, perfect. *Fracture* uneven. *Hardness* 5. *SG* 2.3. *Lustre* vitreous. *Special features* distinctive habit, fuses in a candle flame. *Formation* in cavities in igneous rocks. *Distribution* fine crystals from British Columbia, Nova Scotia and Asbestos (Canada), California and New Jersey (USA), Brazil, Italy, France, Germany, Iceland, Ireland and Skye (Scotland). (L)

Harmotome *Chemistry* $BaAl_2Si_6O_{16}.6H2O$. *System* Monoclinic. *Habit* commonly as twinned cruciform crystals. *Colour* white, grey, brown or yellow. *Streak* white. *Cleavage* one plane.

Autunite

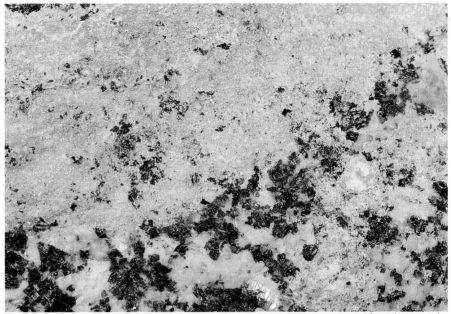

Autunite on granite

Fracture uneven. *Hardness* 4½. *SG* 2.5. *Lustre* vitreous. *Formation* as for other zeolites, in cavities in igneous rocks, especially basalts. *Distribution* fine crystals from New York (USA), Argyll (Scotland) and Andreasburg (Germany). (C)

Heulandite *Chemistry* $(Na,Ca)_{4-6}Al_6(Al,Si)_4Si_{26}O_{72}.24H_2O$. *System* Triclinic. *Habit* tabular, commonly in planar aggregates. *Colour* white, yellow, pink, red, brown or green. *Streak* white. *Cleavage* perfect in one direction. *Fracture* uneven. *Hardness* 3½–4. *SG* 2.2. *Lustre* vitreous or pearly. *Special features* forms pseudomonoclinic crystals. *Formation* with other zeolites in vesicles in basalts. *Distribution* a widespread mineral, good specimens from Iceland, Nova Scotia (Canada), Hawaii, Kongsberg (Norway), Harz (Germany), Italy, India and New Zealand. (C)

Torbernite *Chemistry* $Cu(UO_2)_2(PO_4)_2.8-12H_2O$. *System* Tetragonal. *Habit* tabular, often with square outline, foliaceous. *Colour* emerald green. *Streak* pale green. *Cleavage* perfect basal. *Fracture* uneven. *Hardness* 2½. *SG* 3.3. *Lustre* pearly. *Special features* colour and radioactivity. *Formation* in the alteration zone of veins which contain uranium and copper. In the air torbernite loses water and becomes metatorbernite with $8H_2O$. *Distribution* widespread but uncommon, fine specimens from Mount Painter (Australia), Cornwall (UK), Shinkolobwe (Zaire), North Carolina (USA), Canada and Italy. (C)

Autunite *Chemistry* $Ca(UO_2)_2(PO_4)_2.10-12H_2O$. *System* Tetragonal. *Habit* lamellar with square outline. *Colour* bright yellow or greenish. *Streak* yellow. *Cleavage* perfect basal. *Fracture* uneven. *Hardness* 2½. *SG* 3.2. *Lustre* vitreous or pearly. *Special features* colour and radioactivity. *Formation* in the alteration zones of uranium deposits. Loses some water to become meta-autunite with $2-6H_2O$. *Distribution* well known from Autun (France), Lurisia (Italy), Portugal, Washington and Colorado (USA), Australia and Zaire. (C)

Torbernite

Heulandite

95

Pyromorphite

Wavellite

Wolframite

Turquoise

Apatite

Chrysocolla

Wolframite

Apatite

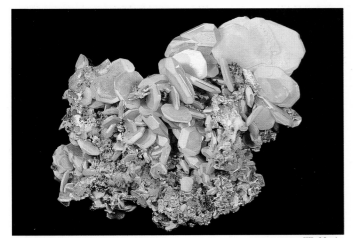

Wulfenite

Vanadinite

Mimetite

Adamite

Pyromorphite *Chemistry* $Pb_5(PO_4)_3Cl$. *System* Hexagonal. *Habit* aggregates of short prismatic crystals; granular or fibrous masses. *Colour* brownish, yellow or dull green. *Streak* yellowish-white. *Cleavage* prismatic. *Fracture* uneven to subconchoidal. *Hardness* 3½–4. *SG* 6.5–7.1. *Lustre* resinous. *Formation* in the oxidized zone of lead veins. *Distribution* fine crystals from Bad Ems and Grube Rosenberg (Germany), Pribram (Czechoslovakia), Berezovsk (USSR), Sardinia, Italy, Mapimi (Mexico), Idaho and Pennsylvania (USA), Broken Hill (Australia) and Cornwall (UK). (K)

Turquoise *Chemistry* $CuAl_6(PO_4)_4(OH)_8.5H_2O$. *System* Triclinic. *Habit* rarely as prismatic crystals, commonly massive, nodular or reniform aggregates. *Colour* light blue to greenish. *Streak* pale green or white. *Cleavage* prismatic. *Fracture* conchoidal. *Hardness* 5–6. *SG* 2.8. *Lustre* waxy, crystals vitreous. *Special features* colour and habit are distinctive. *Formation* in aluminium-rich igneous and sedimentary rocks which have undergone much alteration. *Distribution* fine material from Egypt, Turkestan, Nishapur (Iran); New Mexico, Arizona and Nevada (USA), Chuquicamata (Chile) and the USSR. (K)

Chrysocolla *Chemistry* $CuSiO_3.2H_2O$, plus various impurities. *System* Monoclinic. *Habit* as thin seams and crusts. *Colour* bluish-green. *Streak* white. *Cleavage* none. *Fracture* conchoidal. *Hardness* 2–4. *SG* 2.2. *Lustre* vitreous. *Special features* the colour and habit are distinctive. It is softer than turquoise. *Formation* in the oxidation zones of copper deposits. *Distribution* Italy, Bavaria, Siberia, Arizona and New Mexico (USA), Chuquicamata (Chile), Morocco, Zimbabwe, Adelaide (Australia) and Cornwall (UK). (K)

Apatite *Chemistry* $Ca_5(PO_4)_3(F,Cl,OH)$. *System* Hexagonal. *Habit* short prismatic crystals, also granular and massive. *Colour* greenish to grey, yellow, brown and red. *Streak* white. *Cleavage* very poor, basal. *Fracture* very brittle, uneven or conchoidal. *Hardness* 5 (defines this point on the Mohs scale). *SG* 3.3. *Lustre* subresinous or vitreous. *Special features* hardness and habit. *Formation* a common mineral in many igneous rocks; in metamorphosed limestones and in fossil bones preserved in sediments. *Distribution* very widespread, large crystals in Ontario (Canada), also in Durango (Mexico), Kola peninsula (USSR), Palabora (South Africa), Monte Somma (Italy), Brazil, Burma, Portugal, Austria, Florida and Tennessee (USA). (S)

Wavellite *Chemistry* $Al_3(PO_4)_2(OH)_3.5H_2O$. *System* Orthorhombic. *Habit* typically in radiating spherulitic aggregates. *Colour* pale green, yellow or white. *Streak* white. *Cleavage* perfect prismatic. *Fracture* uneven. *Hardness* 3½–4. *SG* 2.4. *Lustre* vitreous. *Special features* the habit is characteristic. *Formation* commonly found in fractures and joints, on the rock surfaces as a secondary mineral. *Distribution* widespread and not uncommon; large masses in Pennsylvania (USA), Ouro Preto (Brazil) and Cornwall (UK). (S)

Wulfenite *Chemistry* $PbMoO_4$. *System* Tetragonal. *Habit* tabular crystals, often square in outline. *Colour* brownish-yellow. *Streak* white. *Cleavage* pyramidal. *Fracture* conchoidal. *Hardness* 3. *SG* 7.0. *Lustre* adamantine or resinous. *Formation* a secondary mineral which is formed in the oxidation zone of veins containing lead and molybdenum. *Distribution* famous localities include Bleiberg (Austria), Mezica (Yugoslavia), Pribram (Czechoslovakia), Rezbanya (Romania), Tsumeb (Namibia), Arizona, New Mexico and Pennsylvania (USA). (L)

Wolframite *Chemistry* $(Fe,Mn)WO_4$. *System* Monoclinic. *Habit* uncommon crystals are prismatic or tabular; also can be massive. *Colour* dark brown or black. *Streak* dark brown. *Cleavage* one, perfect. *Fracture* uneven. *Hardness* 5–5½. *SG* 7.6. *Lustre* dull to metallic. *Special features* high SG and colour. *Formation* in hydrothermal veins near granite, also in alluvial sands. *Distribution* very fine crystals from Zinnwald and Neudorf (Germany), Fundao (Portugal), Colorado (USA), Bolivia,

Burma (alluvial deposits) and Cornwall (UK). (K)

Mimetite *Chemistry* $Pb_5(AsO_4)_3Cl$. *System* Hexagonal. *Habit* prismatic. *Colour* yellow to brownish. *Streak* white. *Cleavage* none. *Fracture* subconchoidal. *Hardness* 3½. *SG* 7.2. *Lustre* adamantine or resinous. *Special features* high SG, colour, soluble in HCl. *Formation* in the oxidation zone of lead veins. *Distribution* good crystals from Pribram (Czechoslovakia), Freiburg (GDR) and Tsumeb (Namibia). The form called **campylite**, which has curved, barrel-shaped crystals, is known from Cumbria (UK), Pennsylvania and Arizona (USA), Mexico and Italy. (L)

Vanadinite *Chemistry* $Pb_5(VO_4)_3Cl$. *System* Hexagonal. *Habit* prismatic, often hollow crystals; also rounded. *Colour* red, yellowish or brown. *Streak* yellowish-white. *Cleavage* none. *Fracture* conchoidal. *Hardness* 3. *SG* 7.1. *Lustre* resinous or adamantine. *Special features* colour and high SG. *Formation* in the oxidation zone of lead veins. *Distribution* fine crystals at Obir (Austria), Arizona (USA), Argentina, Mexico, Morocco and South Africa. (C)

Adamite *Chemistry* $Zn_2AsO_4(OH)_2$. *System* Orthorhombic. *Habit* commonly as aggregates of small crystals, or botryoidal. *Colour* bright green, yellow or pink. *Streak* white. *Cleavage* good. *Fracture* uneven. *Hardness* 3½. *SG* 4.5. *Lustre* vitreous. *Formation* in the weathered zones of zinc lodes. *Distribution* uncommon, but fine material from Cap-Garonne (France), Laurium (Greece), Mexico, Utah (USA), Chile and Tsumeb (Namibia). (C)

Fossils can occur in great masses

Fossils and their Formation

Fossils are evidence of past life forms as diverse as insects trapped in amber, shells preserved in limestone and footprints left in wet mud by a dinosaur. In order for a fossil to form, a chemical balance has to be maintained between the organic material and the strata in which it becomes trapped. The chance of an organism becoming fossilized is poor, and the fossil record is incomplete, but depending on where the organism lived and what sort of shell or body it had, so its chances of becoming a fossil may improve.

Organisms which live on land exist where weathering and erosion are dominant processes. Even though in some terrestrial environments sediment is deposited in quantity, the remains of creatures and plants which lived on land are comparatively rare. In the sea, however, there is often rapid deposition of sediment which is not removed and which entombs a multitude of organisms, many of which become preserved. Rapid burial of the organism prevents it from being eaten by scavengers or broken by sedimentary processes. An organism with hard parts such as a shell or skeleton, which is also strong enough not to be readily broken, stands the best chance of being preserved. At the other extreme, however, minute creatures such as *Foraminifera* are often preserved in great abundance. In a few well-documented cases soft-bodied creatures have been well preserved. The Burgess shales high in the Rockies of British Columbia, Canada, are unique in containing a wealth of delicate creatures which lived on the Cambrian sea bed and were carried by a series of mud slides to be preserved at the base of a submarine cliff. Another famous example is that of the Solnhofen limestone in Germany, where dragonflies and feathers are preserved in extremely fine-grained sediment.

Whole organisms may occasionally be fossilized, as in the amber on the Baltic coast where insects are trapped in what was originally resin on pine trees. In the southern States of the USA natural pools of tar occur on the land surface. During the Tertiary period mammals stumbled into the sticky swamp and became perfectly preserved, locked away from air and bacteria. Similar processes preserve Arctic mammals trapped in perma-frost in Siberia. There are other examples of such complete preservation but most fossils have undergone change before they are preserved.

The chemical composition of the hard parts of many organisms is such that they can exist in equilibrium with a variety of rocks and environments in the Earth's crust. Calcium carbonate is common in shell-fish and vertebrate skeletons, whilst silica is present in the structure of *Radiolaria* and sponges. The horny nitrogen-bearing carbohydrate, chitin, which occurs in many arthropod exoskeletons, is very stable in a number of geological situations. Nevertheless, changes often take place in the organism, or the fragments of it.

Petrifaction occurs when the organic remains are impregnated with minerals brought in by fluids and solutions seeping through the strata. Such solutions commonly carry calcite, silica and iron minerals, and partial or complete changes can be effected. Though the fossils thus formed may retain their outward shape, the fine detail of the original may be lost. Examples of petrifaction include the iron pyrites shells of molluscs in Jurassic clays and shales, and opal tree stumps from Australia and the USA. The solutions which percolate through the fossil-bearing strata do not only introduce new minerals. Often they remove calcite shells leaving a curved hollow in the sediment. This mould may later become infilled with one of a variety of minerals, thus preserving at least the surface detail and shape of the organic matter. The chemistry of living organisms is based on carbon and many fossils of, for example, plants, fish and graptolites are preserved as a thin black film of this element. After burial in the sediment the more volatile chemicals are released and the carbon percentage thus increases. The outlines of large marine reptiles are sometimes thus preserved, but the best known carbonized fossils are the delicate ferns and other plants from Carboniferous strata.

Fossils can occur without any part of the organism remaining. Footprints, tracks and burrows are trace fossils, indicating where creatures have existed. These can give valuable information as to the size and structure of the animal.

Naming Fossils

It has to be remembered that fossils are biological remains. Therefore they should be named and studied, whenever possible, using similar methods to those a biologist might apply to modern organisms. The system of naming organisms which is today universally applied was established by Linnaeus in 1735. This is hierarchical and divides organisms into groups which have obvious biological similarities. The broadest division is the Kingdom (animals and plants), followed by the Phylum (eg Mollusca), Class (eg Cephalopoda), Order (eg Ammonitida), Family (eg Perisphinctacea), Genus (eg Parkinsonia) and Species (eg parkinsoni). Thus everything has a generic and a specific name as in *Parkinsonia parkinsoni*. Biologists dealing with the modern world usually have the whole organism to study and they can test it in detail to see what its true relations are. Artificial breeding can sometimes be carried out to test the validity of a species. The palaeontological concept of what constitutes a species may be somewhat different from the biological. Many problems can arise when dealing with fossils. Often only a fragment of the original organism survives; evolution has produced, in response to environmental pressures, unrelated creatures living millions of years apart, which look the same; some species

The Petrified Forest, Arizona, USA

exhibit marked sexual dimorphism, which if not recognized will lead to the males and females being named as different species because they are so dissimilar. Many other problems can confuse the palaeontologist, but a group of similar fossils collected from the same horizon can with confidence be assigned to the same species.

Fossils can occur in great masses as in this limestone slab of Silurian age from central England. There are trilobites, bryozoans, brachiopods, coral fragments and crinoids on this piece of the lower Palaeozoic sea bed. (L)

The Petrified Forest, Arizona, USA. Here the fossilized trees have been silicified and turned into opal so that even the growth rings can be seen in section. The trees here, some of which are over 33 m long, did not grow in this place but were carried from a forest some distance away and accumulated as a log-jam among mud and sand during the Triassic period.

Fossils and Stratigraphy

Because evolution produces some creatures which live for only a short period of time, their fossils are found in a relatively short vertical thickness of rock, and they can be very valuable in the relative dating and correlation of strata. A pioneer in the stratigraphic use of fossils was the British canal engineer William Smith. Through

his work he had ready access to new exposures of strata, especially the Jurassic of the south of England. He was quick to realize that certain beds of rock were characterized by particular fossils and that these beds were deposited during the life span of the species. A basic sequence could thus be established with certain beds identified by the fossils they contained and strata some distance away could be correlated with the established sequence if they contained similar fossils. Today a very detailed relative time scale is used, based on these principles. This is divided into time zones (each named after an index fossil), the duration of which may be less than one million years. Accurate palaeontological stratigraphy can be far more precise than radiometric dating. Zones are grouped into larger time units called periods

99

An inland cliff exposure of sandstones, shales and limestones

Exposures of Jurassic strata on the coast of North Yorkshire

A mass of the mollusc Pseudopecten

and a number of periods make up an era.

In order for a fossil to be valuable as a zone fossil it must meet a number of requirements. As mentioned earlier it should have as short a vertical range through the strata as possible. A fossil which has a wide geographical range allows correlation over considerable distances, so those organisms which were free swimming or drifted on sea currents are well suited to being zone fossils. Common, easily recognized fossils with hard fossilizable remains have the edge over such creatures as jellyfish which may be ideal in all other aspects except their chance of becoming fossilized. As well as these criteria, a creature which was not controlled in life by the direct conditions on the sea bed, where sediment was forming, is especially useful as a zone fossil because it may become fossilized in a variety of sediments and therefore allow correlation.

An inland cliff exposure of sandstones, shales and limestones of Carboniferous age, seen here in wintertime, in northern England. These strata contain a variety of fossils including molluscs and plants.

Fossils and Evolution

Our knowledge of evolution is based very largely on the evidence in the fossil record; not only does an overall pattern emerge, but we can also see the detailed development of individual groups. Evolution can be seen not as a steady process, but as a series of rapid expansions punctuated by mass extinctions. Mutation and selection based on environmental influence are important driving forces. There is an amazing diversity of life forms often based on a small number of general types and the further we look back in time the more life forms there are which are remote from the types we recognize around us today. The earliest forms of life were simple soft-bodied organisms and they stood little chance of becoming fossilized. However some of the early fossils, the stromatolites, are mounds of lime secreted by algae and they are not uncommon in many rocks including some of Pre-Cambrian age. At the onset of the Cambrian period there is a relatively sudden abundance of fossils, and this may mark a time when invertebrates were first able to secrete hard shells from carbonates and phosphates. Our knowledge of evolution is thus more detailed for the Phanerozoic part of the time scale (Cambrian to Recent) than for the Pre-Cambrian.

Palaeoecology

Fossil organisms when alive were just as particular about the environmental conditions in which they lived as modern plants and animals. By comparing modern species whose ecology can be studied in detail with similar fossil species it is possible to reconstruct the detailed palaeoecology. Such work is largely based on the principle of uniformitarianism, which has many geological applications. In brief it suggests that the present is the key to the past, and by applying our knowledge of the present geological and biological processes to evidence locked in rock strata we can interpret the past. This all works well up to a point. It must be realized that the older the rocks are, and therefore more remote from us, the more likely it is that conditions may have been very different from those now. The atmosphere, for example, was very different before about 400 million years ago, lacking oxygen which is so essential for many organisms today. Most fossils are the remains of animals and plants which lived in the sea. The vast majority of sedimentary strata are also of marine origin, and it is marine environments that can be thus reconstructed. Continental environments, which often lack fossils, are reconstructed more on sedimentological grounds. In the sea the conditions of salinity, temperature, depth, light and aeration vary and control the organisms which live there. Many organisms are exclusively marine, such as echinoids, many molluscs and the brachiopods; some gastropods and bivalves are restricted to brackish water, others live in fresh water. Plant fossils can give very detailed information about their environment, particularly the climatic conditions. Marine plants tend to live in shallower waters where light can penetrate, and thus creatures which rely on plants for food or cover are restricted to these areas. Land plants are not common as fossils, but when they are found, even as pollen grains or spores, climatic reconstructions are possible. The use of fossils to reconstruct environments and to give a fair indication of the latitude in which rocks were deposited has long been used as firm evidence for continental drift. The many tropical plant fossils in the Carboniferous sediments in North America and Europe, now at high latitudes, have the characteristics of tropical rain forest flora. The stems do not have growth rings, drip leaves are frequent and there is a wealth of species. During Carboniferous times these northern continents were on the equator.

Finding and Collecting Fossils

It is possible to discover where there may be fossils by consulting geological maps and guide books. Drift maps are probably of more use than solid maps because they indicate not only the solid fossil-bearing strata but also the areas which are obscured by river gravel and glacial materials. Sea and inland cliffs, as well as other clean man-made exposures like road cuttings and quarries are often very productive, but care should be taken and permission sought where necessary. It is very important to make notes of what you find and where the locality is and to label specimens accordingly when they are placed in your collection. For detailed scientific work fossil material has to be located to a particular stratum. If a collection is being made simply as a hobby then this information is still of great value as your specimens may one day become part of a museum collection, or be donated to a university department. The wealth of material which has been collected by amateurs and which is housed in cardboard boxes in attics must be quite staggering; if you become tired of your hobby let someone use the material for science.

Geological hammers are valuable for breaking up loose blocks at the foot of a cliff or on the quarry floor, but they are often used very destructively. Collecting fossils should be considered in a similar way to picking wild flowers or collecting butterflies; it is preferably done in moderation. Today excellent cameras can be obtained at relatively low cost, and it is not very difficult to take good photographs of fossils *in situ*. It must always be remembered that amateur palaeontologists make very important discoveries, a classic example being the finding of *Charnia masoni* in the Pre-Cambrian rocks of Leicestershire, UK, by a schoolboy, so an accurate record of where specimens have been seen or found is essential, and the reporting of finds to local museums or other authorities is equally important.

Exposures of Jurassic strata on the coast of North Yorkshire. These are rich in fossils of bivalve molluscs, ammonites, brachiopods and marine reptiles.

A mass of the mollusc Pseudopecten preserved on a shale bedding plane, with much original shell remaining. A specimen from the Jurassic of southern England.

Ketophyllum

Cyathophyllum

Cyathophyllum

Cystiphyllum

Favosites

Halysites

Mesophyllum

Cheddar Gorge, Somerset, UK

Phylum Cnidaria, Class Anthozoa

This class includes the corals and other groups such as the Pennatulacea (eg *Charniodiscus*). The subclasses Rugosa, Tabulata and Zoantharia include all the important fossil corals. Of these the Rugosa and Tabulata are extinct. All groups are marine and first appear in Ordovician strata, although coral-like fossils are known from Cambrian and earlier rocks.

Basic coral morphology consists of a tube- or horn-shaped corallite. This has a cup-like calice on its upper surface in which lives a soft-bodied polyp secreting the calcareous coral. The outside of the corallite, which is either solitary or joined with others in a colony, may be smooth or wrinkled. Inside, the corallite can be a simple hollow tube or complex with radiating vertical divisions (septa), and horizontal dividing structures (tabulae). The walls of the corallite can be thickened by web-like calcareous material called dissepiments. The three main groups of corals share features of this basic design. The Tabulates are the simplest and are generally without septa, although short spinose septa may be present. As their name suggests they have many tabulae. These corals are always colonial and are common fossils in calcareous strata ranging in age from middle Ordovician to upper Permian. Rugose corals are more complex and can be both colonial and solitary. As well as tabulae, the corallites contain numerous septa, inserted in sets of four, and dissepiments thicken the corallite wall. Their geological range is from the Ordovician to the lower Triassic periods. The order Scleractinia (Hexacorals) within the subclass Zoantharia ranges from the middle Triassic to Recent periods and includes the modern reef-building corals. These can be solitary or colonial and have their septa inserted in a six-fold pattern. Dissepiments are present between the septa.

Palaeozoic Corals

Cyathophyllum is a solitary rugose coral with long slender septa and many dissepiments. The main septa nearly reach the centre of the corallite. This genus is from the Devonian strata of Europe, Asia, Australia and North America. (K)

Cystiphyllum is a genus of solitary rugose corals with spinose septa and well-developed dissepiments. The outer wall is heavily ridged. This genus is found in strata of Silurian age, worldwide. (K)

Halysites belongs to the Tabulata and is a colonial genus with a 'chain-like' linking of the individual corallites. Each one has a rounded or oval cross-section. The septa are very short but well-developed; horizontal tabulae are present.

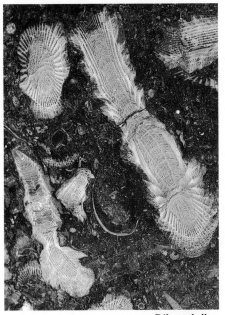

Dibunophyllum

Halysites ranges from middle Ordovician to upper Silurian age and occurs worldwide. (K)

Favosites is a colonial tabulate genus with thin-walled prismatic corallites which are in very close contact. There are numerous tabulae and the walls of the corallites are porous. This genus has a worldwide distribution and occurs in strata of upper Ordovician to middle Devonian age. (K)

Ketophyllum A solitary rugose genus with a moderately deep calice set in a cone-shaped corallite. The outside of the corallite is strongly ridged. Root-like supports are present in well-preserved specimens, at the base of the corallite. Found in Europe and China in strata of Silurian age. (K)

Mesophyllum is a solitary rugose genus with a deep calice. The septa are thin and do not reach the centre. The tabulae and dissepiments are porous; the outer surface of the corallite is strongly ridged. This genus is found in strata of

Thamnopora

Devonian age in Europe, Asia and Australia. (K)

Dibunophyllum is a solitary rugose coral commonly found in lower Carboniferous shallow water limestone strata. The main septa virtually meet the axis, though the minor septa are not continuous. The many small dissepiments form a web-like pattern. This specimen shows both vertical and horizontal sections. The genus occurs in the lower Carboniferous strata of North America, Asia, north Africa and Europe. (C)

Thamnopora A branched tabulate genus with a rounded section. The corallite walls are thick with many pores; the tabulae are thin. This genus is found in strata of Devonian age, worldwide. The specimen is from Torquay, UK. (C)

Cheddar Gorge, Somerset, UK, is cut into limestones of Carboniferous age which contain a fauna of molluscs, corals and brachiopods.

Caninia

Aulophyllum

Syringopora

Clisiophyllum

Lonsdaleia

Lithostrotion

Lonsdaleia

Carboniferous Corals

Clisiophyllum is a solitary rugose genus with a wide axial structure. There are about half as many thin minor septa as there are major septa. This genus is found in lower Carboniferous strata in Europe and Asia. (K)

Lonsdaleia A colonial rugose genus with close corallites which may have a small space between them. The septa are long and the corallite walls strong. The calice is well defined with a central axial structure in each corallite. This genus is found in Carboniferous strata in Europe, Asia, north Africa, Australia and North America. (K)

Caninia is a genus of usually solitary rugose corals which can be cylindrical or conical with long, thick main septa. Tabulae and dissepiments are present in this large genus which is found in strata of Carboniferous age in Europe, North America, north Africa and Australia. (K)

Aulophyllum is a solitary rugose genus with many small septa and dissepiments. There is much detail in the axial area. A Carboniferous genus from Europe, Asia and north Africa. (K)

Lithostrotion A colonial rugose genus with small corallites. The tabulae have cone-shaped centres and the whole mass is often root-like. A genus found in Carboniferous strata in Europe, Asia, north Africa, North America and Australia. (K)

Syringopora is a colonial tabulate genus with long tube-shaped corallites the thick walls of which are joined by small tubular passages. The septa are small and sharply pointed; the tabulae are numerous. With a worldwide distribution, this genus is found in strata of upper Ordovician to Carboniferous age. (K)

Lonsdaleia is a rugose genus. Here a section shows the detail of the septa and dissepiments. (C)

Siphonodendron A rugose coral with a deep calice. There are well-developed septa and dissepiments. This colonial genus is found in Carboniferous strata in Europe. (C)

Mesozoic Corals

Montlivaltia is a solitary scleractinian genus, which grows up to about 10 cm in diameter. There are many straight septa and numerous dissepiments. The corallite is cylindrical or conical with a round cross-section. This genus has a worldwide distribution and is found in strata of lower Jurassic to Cretaceous age. (K)

Montlivaltia

Thecosmilia

Isastraea

Thecosmilia

Trochocyanthus

Thecosmilia A colonial scleractinian genus with branching corallites. The septa are straight and the corallite wall is strengthened with many dissepiments. This genus is a reef-building coral and is known from strata of middle Triassic to Cretaceous age, worldwide. (K)

Trochocyanthus is a genus of small, solitary, cone-shaped scleractinian corals. The outline is circular and there are numerous septa. The calice is moderately deep with a central septa-free area. This genus is found worldwide in strata of middle Jurassic to Recent age. (K)

Isastraea is a large colonial scleractinian genus. There are many six-sided corallites in a colony. These have numerous septa and strengthening dissepiments. Found in Europe, North America and Africa, this genus is of Jurassic to Cretaceous age. (HST)

Thamnastrea is a scleractinian genus with a branching colonial structure. The corallite walls are not well formed, the septa from one corallite merging with those of the next. The detail can be seen clearly in this sectioned specimen. The genus is often a reef builder and ranges from the Triassic to Cretaceous period. The specimen is from the Jurassic strata of Yorkshire, UK. It occurs in Europe, Asia, North and South America. (C)

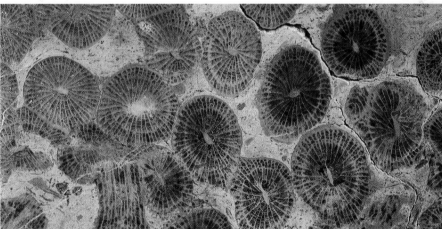

Siphonodendron

Thamnastrea

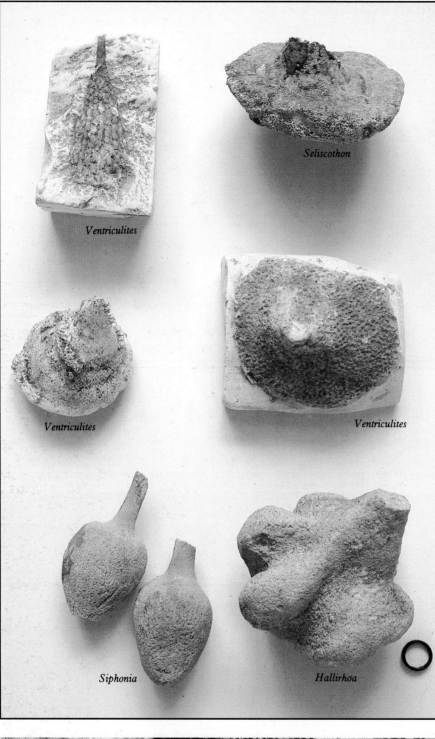

Ventriculites

Seliscothon

Ventriculites

Ventriculites

Siphonia

Hallirhoa

Phylum Porifera

This phylum includes the multicellular sponges, simple organisms with no nervous system. Often they are considered to be between the protozoans and metazoans but are probably not the ancestors of the metazoans. The body is usually upright and like a bag, which may be held on a long stalk. A central opening is present at the top. A variety of skeletons is known; these vary from being jelly-like to siliceous or calcareous. The solid parts of the skeleton, called spicules, are the commonly fossilized material of sponges. They are used for classifying the fossils and have been found in strata dating back to the Cambrian period. Sponges live, today, in all depths of water, and over three-quarters are marine; the remainder dwell in fresh water. They are filter feeders and live on minute organic particles.

Sponges

Ventriculites has a siliceous, thin-walled bowl- or vase-shaped skeleton with linear grooves on its sides. Large pores which join radial canals are present. A genus from the Cretaceous strata of Europe. (K)

Seliscothon is a rhizomorine sponge because of its root-like structure. The skeleton is vase- or funnel-shaped and has a radiate-lamellar structure. A genus from the Cretaceous and Recent strata of Europe. (K)

Siphonia has a 'tulip-shaped' siliceous skeleton attached to the seabed by a stalk. The surface pores are very tiny and are joined to the internal canal system. This genus is found in the Cretaceous and Tertiary strata of Europe. (K)

Hallirhoa has a lobate body structure, which may have a root of varying length. The surface is smooth with minute pores. A genus from the upper Cretaceous strata of Europe. (K)

Raphidonema is a vase-shaped sponge with many small canals permeating its thick walls. It grows up to 5 cm high. The genus has a wide distribution and occurs in Triassic to Cretaceous strata. The specimen is from the Cretaceous rocks of southern England. (C)

Bryozoans

Fenestella Bryozoans are small colonial organisms including the 'sea mats' which are frequently found sticking to algae and stones. Coral reefs often contain abundant bryozoans and their skeletons are very good at binding lime mud, thereby helping the development of reef limestones. They range from Ordovician to Recent age and are common fossils, this genus being one of the most frequent. The specimen is from the Permian strata of north-east England. (C)

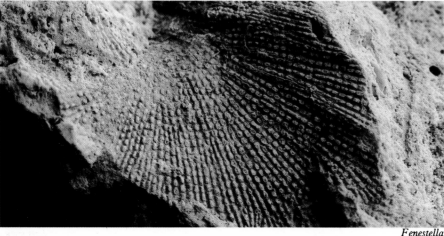

Fenestella

Raphidonema

Solenopora

Stromatolite

Algae

Stromatolite   This is a mound of calcium carbonate secreted by blue-green algae, and such structures are recognized from strata as much as 3,800 million years old. Living stromatolites occur in a number of places, notably Western Australia, where a hypersaline environment in some shallow water coastal areas is too hostile for many organisms and allows the stromatolites to develop unhindered. These algae produce oxygen and their development in the Pre-Cambrian era changed the Earth's atmosphere. The specimen is a small stromatolite, about 15 cm high, and comes from the upper Jurassic strata of southern England. It has been sectioned. (K)

Solenopora   A remarkable red alga from the middle Jurassic strata of Gloucestershire, England. This sectioned specimen shows the alternating layers of pink and white bands. These match variations in the detailed structure of the alga, and though the colour is probably original the banding may be due to subsequent leaching. Rapid burial was probably instrumental in the preservation of the colouring. (HST)

Solenopora   This species has a slender tube-like structure and is often preserved in detail in limestones. It ranges from the Ordovician to Jurassic periods; the specimen is from the Ordovician strata of southern Norway. (K)

Solenopora

Asterophyllites

Annularia

Pecopteris

Mariopteris

Alethopteris

Neuropteris

Neuropteris

Neuropteris preserved in an iron-rich nodule

Vascular Plants

The fossil record contains far fewer remains of plants than animals. Various reasons account for this; plants contain much soft tissue and the more primitive plants have no hard, easily fossilizable material; the evolution of plants is such that the larger, tougher-stemmed genera do not develop until well into the Palaeozoic era, whereas many invertebrate organisms with shells are common as early as the lower Cambrian period. Plants live in almost every habitat today and during the past many types have colonized the land. Strata laid down on the continents are, in favourable conditions, preserved to become part of the geological record and these may contain plant fossils. However, often mud and sand deposited on the land surfaces will be weathered and eroded before it becomes rock, and so any fossil material contained in it will be lost.

Many of the most prolific assemblages of plant fossils are those deposited in deltaic environments, for example those of the upper Carboniferous and Jurassic periods, and the swampy conditions of the early Tertiary period. Such conditions allow the preservation of soft tissue because the stagnant, oxygen-poor environment prevented the normal rapid decay of the plants which would have taken place in the air.

Plant fossils are usually found in the form of carbon impressions but replacement with calcite and quartz (sometimes as opal) can provide excellent three-dimensional material. Silicified plant fossils allow detail of the cell structure to be examined. Most plant fossils, however, because of the delicate nature of plants and the way many live, with roots in the ground and fragile stems and leaves above, tend to be of the different parts, separated and fragmented. The names of these various parts of the same plant may therefore be different; the large Carboniferous lycopod trunk called *Lepidodendron* has roots named *Stigmaria*, while the cones are called *Lepidostrobus*.

Plant Evolution

Among the earliest organisms to be found as fossils are the limy mounds of stromatolites secreted and built by blue-green algae. They are relatively common fossils from Pre-Cambrian strata, the oldest being as much as 3,500 million years old, and occur in Canada, Australia and southern Africa. Stromatolites are also found in rocks throughout the geological record, and live today, a famous site being Shark Bay, Western Australia.

Such algae are the ancestors of the vegetation that today covers much of the Earth's surface. An important result of the development of these algae, and subsequent plants, is that they changed the Earth's primitive anoxic atmosphere, by photosynthesis, into one with abundant oxygen. The fossil record of the early, simple plants is very poor, but in the Silurian rocks of Wales stems and branches of *Cooksonia* are found. This genus carried spore capsules at the end of rigid stems. Microscopic spores are found in earlier strata, but *Cooksonia* is one of the earliest definite plant fossils. Such plants would rely on a very moist habitat for their reproduction.

During the Devonian period the fossil record gradually becomes richer in plant remains and a famous, well-documented fossil flora is that from the Rhynie chert in Aberdeen, Scotland, where a swamp community is preserved in great detail. The preservation is such that cell structure can be examined microscopically. Devonian land plants were small, a 'forest' being only about a metre high. Towards the end of the Devonian period ferns and seed-bearing plants appear, and in the Carboniferous period luxuriant forests of giant lycopods and horsetails spread across many of the low-lying delta and swamp regions in what is now North America, Eurasia and Siberia. A very rich flora is recorded from these upper Carboniferous deposits including the lycopods *Lepidodendron* and *Sigillaria*, and the horsetail *Calamites*; seed-ferns include *Neuropteris* and *Alethopteris* and the gymnosperms (which reproduce by seed) are represented by *Cordaites*.

In Permian and Triassic times the distribution of land plants changed in that much of the northern hemisphere was arid and the Pteridophytes (ferns, clubmosses and horsetails) became fewer, while the conifers, cycads and ginkgos developed. Strata of this age, formed on the continental mass called Gondwanaland (made up of India, South America, southern Africa, Australasia and Antarctica), contain a flora which is famous for the seed-fern *Glossopteris*, used as evidence of continental drift.

In the Jurassic period the Cycadophytes, which have rough columnar stems covered with the previous seasons' leaf bases and topped with feathery leaves, predominated. In such regions as Siberia and China there are good fossil floras and the deltaic strata in north-east England contain excellent examples of the common Bennettitales and Cycads. *Williamsonia* is a well-known Bennittitale from sediments formed in swamp conditions.

Conifers became widespread during the Mesozoic. This well-known group with their thin needle-like leaves secrete sticky resin in which insects become trapped, possibly to be preserved when the resin hardens to amber. Plants were without true flowers until the angiosperms (which are predominant today) developed towards the end of the Jurassic period. Flowers provide a means of sexual reproduction allowing new gene combinations to occur, and these plants go to great lengths to achieve cross-pollination. Some use scents, nectar and colours to attract insects; others produce pollen which will be carried by the wind. Fossil angiosperms are well known in the Tertiary strata of south-east England and Mesozoic and Tertiary rocks in North America, Europe, the USSR and the Far East. Because many angiosperm genera which are fossilized still occur today, climatic reconstructions are possible. This is especially important for sediments deposited during the Quarternary glaciation. Pollen grains found in ice-age sediments allow accurate reconstructions of ice advances and retreats to be made.

Asterophyllites is a genus of delicate articulate plants with a slender central stem and equally spaced nodes bearing slender leaves. It occurs in the upper Carboniferous strata of Europe. (L)

Pecopteris is a fern with many small fronds joined to the stem. This specimen is one of the side shoots from the main stem. Some species of this genus were seed ferns, others bore spore capsules and were true ferns. The genus occurs in upper Carboniferous strata in Europe, North America and Asia. (S)

Annularia is the generic name given to the leaves of the horsetail *Calamites*. These leaves are in delicate rosettes. The plant bore its spore cases in cones on the tips of the branches. This genus is from Carboniferous strata of the USA, Europe, Canada and China. (S)

Mariopteris is a genus of Pterophytes (true ferns) with fronds which are much divided. Spores are the means of reproduction and moisture is needed for fertilization. A genus from Carboniferous strata of Europe and North America. (L)

Alethopteris is a genus of Pteridosperms (seed-ferns). It is about the same size as *Neuropteris* and has a similar structure, though its pinnules are more pointed. This genus is from Carboniferous strata of Europe and North America. (L)

Neuropteris is another Pteridosperm with a frond usually found fossilized away from the main plant. The pinnules are oval and commonly preserved as carbon impressions. The Pteridosperms are an extinct group but were common in Carboniferous delta swamps. The specimens are from the coal measures of Staffordshire, UK; it occurs in the upper Carboniferous strata of North America and Europe. (S)

Neuropteris preserved in an iron-rich nodule. The specimen is from Illinois, USA. (JM)

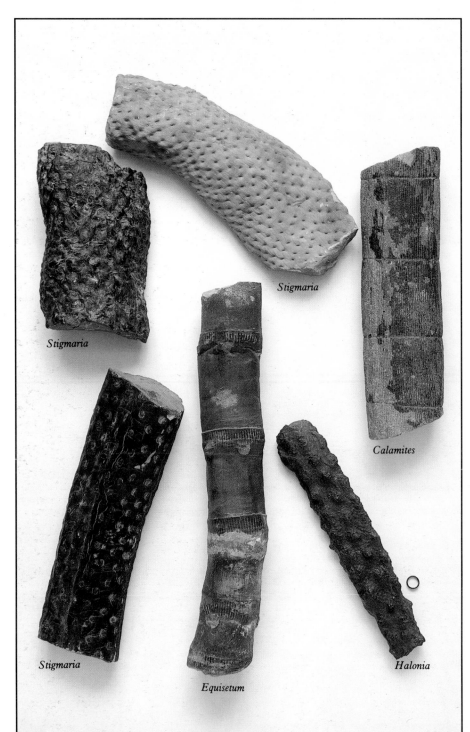

Stigmaria

Stigmaria

Calamites

Stigmaria

Equisetum

Halonia

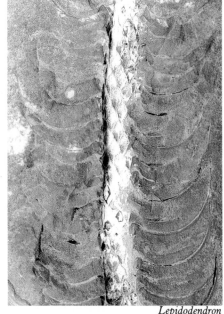

Lepidodendron

Stigmaria is the generic name for the fossilized roots of the giant clubmoss *Lepidodendron*. These rhizomes are sturdy, thick and cylindrical, though (as in these examples) they are often crushed during fossilization. The scars represent the points where smaller slender rootlets left the main root. The whole root system extended almost horizontally from the plant. It is found worldwide in strata of Carboniferous and Permian age. (S)

Calamites is a genus of horsetails with a typically jointed stem. It grew to 20 m and is common in rocks of Carboniferous and Permian age in the USSR, USA, China, Korea and Europe. (CP)

Equisetum is another horsetail genus. It is represented today by small species but in the past grew to considerable size, some being over 40 m high. They are characterized by a hollow stem with equally spaced joints where needle-like leaves occur. There are vertical ribs on the stem and branching underground rhizomes. The genus lives in damp conditions and is found worldwide in strata from Devonian to Recent age. The specimen is of Jurassic age from North Yorkshire, UK. (L)

Halonia is the generic name for a fossil root system characterized by rows of large nodes and a circular cross section. It is found in the Carboniferous strata of Europe. (L)

Cooksonia is the first genus of vascular plants to be found in the fossil record. The earliest examples are from the Silurian strata of Wales. The example shown is from Scottish Devonian strata. The stems branch and carry spore capsules on their upper tips, and contain water-conducting cells of xylem. The genus belongs to a group of plants called Psilophytes and has been found fossilized in North America, Europe, Africa, Asia and Antarctica. The specimen is 3 cm across. (RMS)

Lepidodendron Two specimens of this common Lycopod (clubmoss) from the Carboniferous period. One shows the leafy branch and the other a detail of the scarred stem. (D)

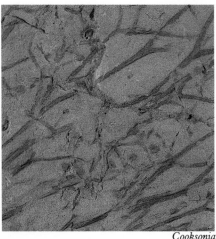

Cooksonia

Lepidodendron

Williamsonia

Williamsonia

Coniopteris

Glossopteris

Ginkgo

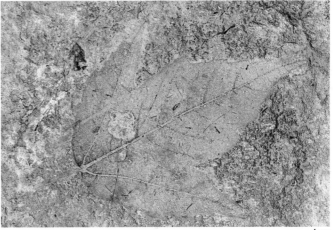

Acer

Coniopteris is a delicate Mesozoic fern which has the frond divided two or more times, and a short stalk. It has a widespread distribution, being found in Europe, Alaska, Poland, India, Japan and the USSR. (JM)

Glossopteris is a genus of deciduous Pteridosperms which lived as a bushy plant some 6 m high. The leaves, shown here, are commonly preserved as fossils in strata of Permian and Triassic age in South Africa, South America, Antarctica, India and Australia. That these regions were once joined in a supercontinent called Gondwanaland is to some extent proved by the distribution of *Glossopteris*

and other fossils. This specimen is from Adamstown in Australia. (C)

Williamsonia is classified with the Cycadophytes, the most ancient group of gymnosperms. This genus belongs to the Bennettitales which had hermaphroditic 'flowers'. It is characterized by pinnate leaves, as on these specimens. The range of the Bennettitales is Triassic to upper Cretaceous, with a worldwide distribution. *Williamsonia* is a Jurassic genus. (C, JM)

Ginkgo is a gymnosperm which is still represented by a single species living today in

China, and much grown by horticulturalists. The leaves have a distinctive fan shape with veins. The tree is deciduous and reaches about 30 m high. There are separate male and female plants. The leaves are not uncommon as fossils in strata from Permian to Recent age, and are found worldwide. The specimen shows leaves which are 3 cm across from the Jurassic strata of northern England. (C)

Acer A genus of modern plants, this delicate leaf is preserved in fine-grained limestone of Miocene age from northern France. (L)

Cenoceras

Orthoceras

Actinoceras

Pseudocenoceras

Dawsonoceras

Cenoceras

Gomphoceras

Phylum Mollusca

This is a very diverse phylum including shelled and unshelled forms. There is a soft body and in many groups a muscular 'foot'. The detailed structure of the body varies tremendously according to where the particular group lives. Some have large bodies with tentacles, eyes and highly organized nervous systems; some have a twisted body enclosed in a spiral shell; some have a more simple body and live under a single, quite flat shell; others live between two valves and are relatively sophisticated but lack eyes and tentacles. The majority live in the sea, but there are also fresh and brackish-water dwellers and a number of forms live on dry land and even climb trees. Their size varies from enormous to microscopic.

Class Cephalopoda

These are marine creatures which are shelled and capable of swimming freely. They have great diversity of form and habits and evolved rapidly. Movement is by jet propulsion, a current of water being squirted out of a funnel among the tentacles; these may also help during locomotion near or on the sea bed. The body of extinct forms was probably not unlike that of squids and cuttles. The shell is a masterpiece of design. It is coiled in a flat spiral, the coils or whorls becoming smaller away from the aperture. If the shell were to be uncoiled it would be like a tapering cornet with a groove along one side for the curved ventral side to rest in when coiled. Internally the shell is divided into chambers separated by septa. Near the aperture is a large body chamber and behind this are the buoyancy chambers. These are linked by a thin tube, the siphuncle, which helps to regulate the density of the fluid in them. In the Nautiloids the siphuncle is central or nearly so, and in the Ammonoids the siphuncle is marginal, very near the ventral surface of the shell. Where the septa join the inner surface of the shell wall there is a suture line which is most complex in the mesozoic ammonites. Earlier Ammonoids and the Nautiloids have more simple sutures. The suture lines can only be seen when the outermost surface of the shell has been removed and these lines may be of diagnostic value. Shell shape varies greatly in the Cephalopods. Some have shells which coil in such a way as to obscure the inner whorls (involute coiling); others coil with hardly any overlapping (evolute coiling); other groups have shells which are straight or uncoiled. The tentacled body has a good nervous system and eyes, and many species (both modern and fossil) have an ink sac for squirting defensive fluids. The class is the most sophisticated of the invertebrates. Cephalopods first appear in the fossil record in the Cambrian period and the class is represented today, though many groups which have been numerous in the past are now extinct.

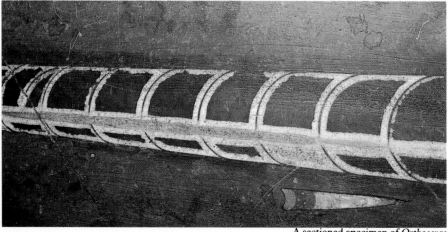

A sectioned specimen of Orthoceras

Nautilus

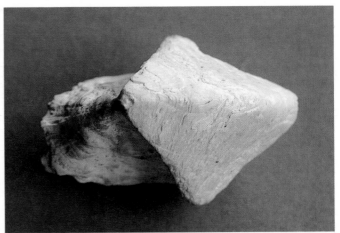

Rhyncholite

Sub-class Nautiloidea

First appearing in the Cambrian period, this group becomes greatly diversified during the Ordovician period and is still represented today. In many early forms the shell is straight (orthoconic), but coiled shells are a characteristic feature of the sub-class. Large straight-shelled Nautiloids were fierce predators on the Palaeozoic sea bed. In Nautiloids the siphuncle is central and the sutures are simple curves. Only very few species of *Nautilus* survive today; typical among them is *Nautilus pompilius*, which lives around Fiji, the Philippines, eastern Australia and New Guinea. It is interesting to note that dead shells drift on ocean currents for great distances, indicating that chambered molluscs may have a greater distribution in sediments than their own habitat.

Actinoceras is an orthoconic Nautiloid which can grow to considerable size. A characteristic feature is the bulbous swelling of the siphuncle between the septa. The siphuncle has a central tube with radiating off-shoots. The genus is frequently found in strata ranging from lower Ordovician to upper Carboniferous age worldwide. The specimen is from the Ordovician strata of north-east Greenland. (K)

Dawsonoceras grows to considerable size but typically is about 12 cm long. The shell has a circular cross-section and is covered with well-spaced, stout ribs. The genus is from strata of Silurian age, in Europe, Asia, North America and Australia. (K)

Orthoceras grows up to several metres long. The shell is almost cylindrical and increases in diameter lengthwise. In this specimen the septa are clearly seen. The genus has a worldwide distribution in rocks of lower Ordovician to Triassic age; the specimen shown is from Devonian strata from the Atlas Mountains, Morocco. (K)

Gomphoceras has an average length of about 10 cm. It is a strange egg-shaped shell which has broad ribs over about half its length and a slit in the ventral surface. Its true relationships with other Cephalopoda are uncertain. The specimen is from the Silurian strata of Shropshire, UK, and the genus ranges across Europe in strata of this age. (K)

Dentalium

Cenoceras can have an evolute to involute shell with a globose outline. There are growth lines and thin longitudinal lines on the shell. The sutures are shallowly lobed and the siphuncle has a variable position but it is never very close to the dorsal or ventral surfaces. It is found worldwide in strata of upper Triassic to middle Jurassic age. (K)

Pseudocenoceras has an involute shell with an almost rectangular cross-section. The umbilical shoulders are rounded and the siphuncle almost central. This genus is found in the Cretaceous strata of Europe and north Africa. (K)

Nautilus A cut and polished specimen from strata of lower Jurassic age, showing the chambers and central siphuncle. This genus still lives in Indo-Pacific waters. (D)

A sectioned specimen of Orthoceras A straight-shelled nautiloid from Shanghai, China. The sub-central siphuncle and septa are made of pale calcite. (D)

Rhyncholite This unusual specimen is of the beak-like jaw mechanism of a Cretaceous nautiloid from the chalk of Norfolk, UK. (HST)

Class Scaphopoda

These have a very thin tapering shell which is tubular and open at both ends. The ventral surface can be convex. When alive, the anterior end is stuck into the sediment surface and the tapered posterior part protrudes into the water. The organism has a head, radula and a tentacle-like feeding organ. They are more common today than in the past and are known from strata as old as Ordovician.

Dentalium is a common form which often has longitudinal ribs. The apex may be polygonal and it grows to a maximum of some 15 cm. The three forms shown are from left to right, *Dentalium priseum* from the Carboniferous strata of Dumfries, UK, *D. sexangulare* from the Pliocene strata of Tuscany, Italy, and *D. striatum* from Eocene strata in Hampshire, UK. (K)

Reticuloceras

Cyrtoclymenia

Cymaclymenia

Joannites

Arcestes

Clymenia

Sudeticeras

Cyrtoclymenia

Prolecanites

Goniatites

Gastrioceras

Cladiscites

Sub-class Ammonoidea

Creatures from this group are first known from Devonian strata and they develop rapidly throughout the Upper Palaeozoic and the Mesozoic. By the end of the Cretaceous period they were extinct. Their development through time into many different genera and species and their widespread geographical distribution (they were free-swimming with a buoyant shell) make the Ammonoids excellent time discriminators or 'zone fossils'. The stratigraphic subdivision possible using them is so refined that an ammonite horizon is normally of a duration of 100,000 years, and zones and sub-zones may be as little as 250,000 to 1 million years.

Their nearest living relatives are the squids and octopus, though the chambered shell of Ammonoids resembles that of the more distantly related *Nautilus*. Important differences between the Ammonoids and *Nautilus* include the position of the siphuncle (marginal in Ammonoids, central in *Nautilus*) and the more complex suture line of the Ammonoids. This is especially noticeable in the ammonites which had a system of 'fan-vaulting'. There the septa meet the shell giving a very complex suture.

Because they developed rapidly into many different species, homeomorphic similarities arise between unrelated groups. Sexual dimorphism is widely recognized within the sub-class in Jurassic examples. This is almost always expressed in significant shell size differences between supposed males and females of a species, though to be certain as to which sex is which is not easy with fossils, so the terms microconch (m) and macroconch (M) are used when sexual dimorphism is recognized. For this diagnosis to be accurate mature complete shells have to be considered. Complete shells will have the body chamber, recognized by the absence of suture lines, present. Maturity can be ascertained by the way sutures tend to crowd together near the body chamber in mature ammonites; slight uncoiling of the seam between the whorls near the aperture and a change in the ornamentation and sculpture around the aperture.

The term Goniatites is used in general for the Ammonoids of the Devonian and Carboniferous periods, Ceratites for those of the Permian and Triassic and Ammonites for the genera found in Jurassic and Cretaceous rocks. These are only general terms and the detailed evolution of the Ammonoidea involves very many groups at sub-order and other levels.

Palaeozoic and Mesozoic Ammonoids

Gastrioceras can have a diameter of up to 10 cm and has involute coiling of the shell, which is round. The edge of the umbilical area has nodes which almost develop into ribs. The specimen is from the upper Carboniferous strata of Lancashire, UK, and the genus is found in strata of this age in Europe, Asia, North America and north Africa. (K)

Reticuloceras is a genus with a shell diameter of only about 3 cm. The shell has involute coiling with widely spaced ribs. The shell cross-section is very rounded. This genus is found in strata of upper Carboniferous age in Europe, Asia and North America. (K)

Goniatites grows to a diameter of about 5 cm and has a globular shell with involute coiling. The suture lines have pointed lobes and rounded

Two specimens of *Goniatites*

saddles. Some species have spiral ornamentation. This specimen is from the Carboniferous strata of the Isle of Man, UK, and the genus is found in Europe, Asia, North America and north Africa. (K)

Prolecanites grows to a diameter of 20 cm. The coiling is slightly involute, and the whorls somewhat flattened. The characteristic suture lines are easily visible on this specimen. The genus is found in strata of lower Carboniferous age in Europe, Asia and North America. (K)

Cyrtoclymenia has involute coiling. The sutures are simple and there is a little ornamentation with faint ribs. The specimens are from the Devonian strata of Poland and the genus is found in strata of this age in Europe, north Africa and Western Australia. (HST)

Clymenia grows to about 8 cm in diameter, with evolute coiling and a generally smooth shell with faint ribs. The specimen is from the upper Devonian strata of Poland and the genus is found in Devonian strata in Europe, Asia and north Africa. (HST)

Cymaclymenia has a subinvolute shell with growth-lines. The umbilical margin is steep and the umbilicus quite deeply set. The specimen is from Devonian strata in Poland and the genus has a worldwide distribution, except for South America and Antarctica. (HST)

Sudeticeras has a small involute shell with a globular outline and no distinct ornamentation, apart from fine ribs. The specimen is from the lower Carboniferous strata of Yorkshire, UK, and the genus is found throughout Europe, North America, north Africa and Eurasia. (K)

Arcestes reaches about 10 cm diameter with an involute shell. The outline is bulbous and the shell is smooth. Suture lines are complex. The specimen is from Triassic strata in Austria; the genus is found worldwide, with the exception of the UK. (K)

Cladiscites grows to a large size and is involute with the inner whorls being obscured. The shell becomes wider towards the venter. The specimen illustrated has had the outermost layer of shell worn off and the complex suture lines are well shown. The specimen is from the Triassic rocks of Austria and the genus is found in strata of this age in Europe (though not the UK), the Himalayas and Alaska. (K)

Joannites has a medium to large globular shell with such involute coiling that the inner whorls are obscured. The shell is smooth. The specimen is from the Triassic strata of Austria and the genus is found in Europe (though not the UK), the Himalayas and North America. (K)

Two specimens of Goniatites from the Carboniferous strata of Yorkshire, UK, showing the characteristic globular shell with involute whorls. The sectioned specimen has infilled chambers and a marginal siphuncle. (C)

The Durness limestone of north-west Scotland. One of the oldest formations to contain fossils of molluscs, this limestone ranges in age from Cambrian to Ordovician and is seen here as pale rock exposed on a dipping thrust plane.

Anaptychus   A specimen of the lower Jurassic ammonite *Arnioceras* showing the *Anaptychus* (lower jaw) *in situ* near the aperture of the shell. (JM)

The Durness limestone of north-west Scotland

Anaptychus

Liparoceras

Promicroceras

Aegoceras

Aegoceras

Pleuroceras

Microderoceras

Psiloceras

Oxynoticeras

Eoderoceras

Asteroceras

Pleuroceras

Amaltheus

Echioceras

Lower Jurassic Ammonites

Eoderoceras is a genus with an average diameter of about 10 cm. The shell has evolute coiling, all the whorls being visible. There are well-spaced, broad ribs with nodes on the inner whorls. The genus is from lower Jurassic strata and is found worldwide. (K)

Echioceras has a diameter of up to about 10 cm, with an evolute shell and open umbilicus. There are coarse, well-spaced ribs. The genus is found in lower Jurassic strata worldwide. (K)

Pleuroceras grows to a diameter of about 10 cm. The shell is sub-evolute and the whorl section is rectangular. There are strong, well-spaced ribs which may have tubercles and even spines on them. The venter is flat with an obvious keel. The dark specimen is from the lower Jurassic strata of the Yorkshire coast, UK, the other specimen is from Germany. The genus is from lower Jurassic strata in Europe and north Africa. (CP, K)

Amaltheus may have a diameter of up to 8 cm with involute coiling. The shell is very compressed. The ribs bi- and trifurcate near the venter; the keel is corded. The genus is from lower Jurassic strata of Europe, north Africa, Asia and North America. (K)

Microderoceras can have a diameter as large as 20 cm. The coiling is evolute and the shell has coarse ribs and weak spines. Across the venter the ribs are weaker. The genus is found in lower Jurassic strata in Europe, Central and South America. (CP)

Asteroceras has a diameter up to about 10 cm, with a sub-evolute shell. The size of the whorls increases quickly from the umbilicus and the strong ribs curve forward as they reach the venter. The keel has grooves on each side. This genus is found in lower Jurassic strata in Europe, North America and Asia. (K)

Aegoceras (m) is a small- to medium-sized genus with a diameter of about 6 cm. The shell has evolute coiling and large stout ribs, which curve forward at the venter. Weak tubercles may occur on the ribs. This genus is thought to be the microconch in a sexually dimorphic pair with *Liparoceras*. The genus is found in lower Jurassic rocks in Europe. (K, CP)

Liparoceras (M) grows to a diameter of about 10 cm and has a stout involute shell with rounded whorls and a deep umbilicus. The ribs are strong and there are spines near the venter. The genus is from lower Jurassic strata of Europe, north Africa and Indonesia. (CP)

Psiloceras grows to about 7 cm maximum and has an evolute shell devoid of ornamentation except for the ribs. It is very often found as a crushed impression and is the earliest zonal ammonite in the Jurassic strata. This genus comes from Europe, Asia, North and South America. (K)

Promicroceras is a genus which grows to a diameter of about 3 cm. The whorls are evolute with an almost circular whorl section. The sharp ribs become flat across the venter. This genus is found in the lower Jurassic strata of Europe. (CP)

Oxynoticeras reaches 8 cm in diameter and has a flattened shell with involute coiling. The ribs

are most apparent on the umbilical margin. The genus is from lower Jurassic strata in Europe, north Africa, Japan, Indonesia and South America. (K)

Promicroceras preserved on a single bedding plane. These specimens are from the lower Jurassic strata of Somerset, UK. In the past, strata such as these were cut and polished for ornamental purposes. Some of the specimens have original shell remaining, others show internal moulds. (HST)

A wave-cut platform, below cliffs of alternating shales and limestones, in Somerset, UK, containing fossils of large ammonites of Jurassic age.

A wave-cut platform

Promicroceras preserved on a single bedding plane

Hildoceras

Tulites

Chondroceras

Leioceras

Morrisiceras

Parkinsonia

Graphoceras

Procerites

Harpoceras

Ludwigia

Stephanoceras

Dactylioceras

Dactylioceras

Hildoceras

Ludwigia and *Ludwigina*

Procerites and *Siemiradzkia*

Brasilia and *Ludwigella*

Lower and Middle Jurassic Ammonites

Stephanoceras (M) may grow to a diameter of 15 cm and has an evolute shell with relatively rounded whorl cross-section. The ribs are coarse on the umbilical margin but bi- and trifurcate across the venter from a small tubercle. The specimen is from Dorset, UK, and the genus is found in middle Jurassic strata worldwide. (K)

Graphoceras can have a diameter of up to 8 cm and has involute coiling with the umbilicus largely obscured. There are weak sickle-shaped ribs which bifurcate across the venter and the whorl section is flattened. The umbilical shoulder is steep and the venter sharp. The genus is found in middle Jurassic strata in Europe, Asia and Africa. (K)

Chondroceras is a small bulbous genus reaching about 4 cm in diameter. It is involute with a rounded whorl section. The faint ribs bi- and trifurcate across the venter. This genus is from middle Jurassic strata in Europe, north Africa, New Guinea, Indonesia, North and South America. (K)

Procerites (M) can grow to a very large size. The coiling is sub-evolute and the whorl section is moderately flattened. There are strong ribs which bifurcate across the venter. The genus is found in middle Jurassic strata in Europe. (K)

Tulites (M) grows to about 15 cm diameter and has a heavy rotund shell wtih involute coiling and a very deep umbilicus. There are weak ribs which may have nodes on the umbilical margin. The specimen is from Somerset, UK, and the genus is found in middle Jurassic strata in Europe and Saudi Arabia. (K)

Ludwigia (M) has a medium to large shell which can be 12 cm in diameter. The whorls have involute coiling with only a small part of the inner whorls visible. The umbilical margin is steep and the sickle-shaped ribs bifurcate halfway between the umbilical margin and the venter. The ventral keel is sharp. The genus is from middle Jurassic strata in Europe, north Africa, Iran, Siberia and South America. (K)

Morrisiceras (M) has a diameter of about 8 cm and the shell has a rounded outline, with a deep umbilicus. The coiling is involute. The inner whorls have bifurcating ribs, the outer whorl has ribs across the venter. The genus is found in middle Jurassic strata in Europe. (HST)

Leioceras has an involute shell with a very flat whorl cross-section and a sharp keel. There are numerous sickle-shaped ribs. The specimen is from Germany and the genus is found in middle Jurassic strata in Europe, north Africa, Iran and the Caucasus. (CP)

Hildoceras may reach 12 cm or more in diameter and has a shell with evolute coiling and a strong median groove on the whorls. The ribs are sickle-shaped, and the keel has a furrow on each side. Both the specimens are from the UK: the one with its original shell is from Northampton, the darker one from Whitby. The genus is found in lower Jurassic strata in Europe, Asia Minor and Japan. (HST, K)

Harpoceras is a medium to large genus which may be over 20 cm in diameter. The shell has involute coiling and is compressed with a sharp ventral keel. The ribs are sickle-shaped. The genus has a worldwide distribution in strata of lower Jurassic age. (K)

Dactylioceras has a diameter of about 5 to 10 cm and evolute coiling. The whorl section is sub-circular. There are strong ribs which bifurcate across the venter; there is no keel. Tubercles may be present on the inner whorls. The genus is common in lower Jurassic strata worldwide. (CP, K)

Parkinsonia grows to about 15 cm in diameter and has an evolute shell. The whorl cross-section is compressed and the ribs bifurcate at the venter, but are interrupted by a smooth band around the venter. The genus is found in middle Jurassic strata in Europe, north Africa, the USSR and Iran. (K)

Dimorphism in Ammonites

Sexual dimorphism is commonly expressed by considerable size differences between one sex and the other. It is impossible to be certain as to which is male and female when dealing with fossils. The larger shell is called the macroconch (M); the smaller one the microconch (m). There are definite ways of telling that the shells are mature and complete and that the smaller ones are not juveniles or simply the innermost whorls from fragmented macro-conchs. Three examples are illustrated:

Ludwigia (M) and **Ludwigina (m)**, from the middle Jurassic strata of Skye, Scotland. (HST)

Procerites (M) and **Siemiradzkia(m)**, from the middle Jurassic strata of Cape Mondego, Spain. (HST)

Brasilia (M) and **Ludwigella (m)**, from the middle Jurassic strata of Dorset, UK. (K)

Baculites

Hamites

Kosmoceras

Laevaptychus

Euhoplites

Scaphites

Cardioceras

Anahoplites

Kosmoceras

Peltoceras

Pectinatites

Quenstedtoceras

Schloenbachia

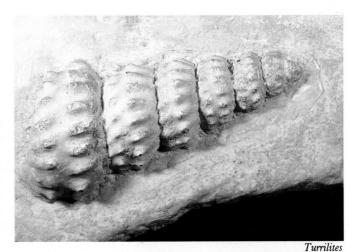

Pavlovia

Turrilites

Middle and Upper Jurassic/Cretaceous Ammonites

Scaphites has a maximum diameter of about 10 cm and has a shell with broad, flattened whorls. The inner whorls are tightly coiled and the outermost whorls become detached. There are many fine ribs and tubercles on the umbilical and ventral margins. The specimen is from South Dakota, USA, and the genus is found in Cretaceous strata in Europe, South Africa, Canada, USA, Chile and Australia. (K)

Euhoplites has a diameter of about 5 cm when mature. The shell is sub-evolute with a deep umbilicus. A deep groove runs along the venter. Very strong ribs divide from nodes and curve forwards as they cross the venter. The specimen is from Kent, UK, and the genus is found in lower Cretaceous strata in northern Europe, Greenland and Alaska. (K)

Anahoplites can be up to 10 cm in diameter and has involute coiling and a compressed shell. The venter is squared and the shell is smooth with very weak ribs; nodes are present on the umbilical margin. The genus is found in lower Cretaceous rocks in Europe. (K)

Schloenbachia reaches a diameter of about 8 cm and has relatively involute coiling. There are strong curved ribs with tubercles. The venter is smooth and has a thin keel. The genus is found in Cretaceous strata in Europe, Greenland and south-west Asia. (CP)

Peltoceras (M) is a large stout genus which reaches over 20 cm in diameter. The coiling is evolute and there are massive ribs with tubercles on the outer whorls. The genus is found in upper Jurassic strata, worldwide. (K)

Pectinatites (M) can be over 12 cm in diameter. The whorls have sub-evolute coiling and the strong ribs bi- and trifurcate near the venter. The specimen is from Oxfordshire, UK, and the genus is found in upper Jurassic strata in Europe, the USSR and Greenland. (K)

Baculites grows up to 2 m long and is an uncoiled form which has a minute coiled early stage. The shell is straight or slightly curved and is commonly found as broken pieces. There are ribs and tubercles on the venter and the aperture has a dorsal extension or rostrum. The specimen is from South Dakota, USA, and the genus is found in upper Cretaceous strata, worldwide. (K)

Hamites is an uncoiled genus with the first part of the shell openly coiled. There are two or three almost straight uncoiled sections which are joined by sharp curves. The cross-section is circular or oval and strong ribs encircle the shell. The genus is found in lower Cretaceous strata in Europe, Asia, North America and Africa. (K)

Kosmoceras may reach 6 cm in diameter. The coiling is involute and there are nodes and spines on the ribs. The venter is flat. This genus provides a classic example of sexual dimorphism; the small microconch is illustrated showing the typical lappet extending from the aperture. The large form, the macroconch, is simpler and without a lappet. The genus is from the middle Jurassic strata and is found worldwide. (K)

Quenstedtoceras (m) grows to about 6 cm in the microconch form, which is the one illustrated. The coiling is sub-evolute and the strong ribs are sickle-shaped. Complete specimens have a rostrum at the aperture. The macroconch is much larger and has less ornament. The genus is from middle and upper Jurassic strata, and is found worldwide. (K)

Cardioceras may be 5 cm in diameter and has sub-evolute to involute coiling. The strong ribs bifurcate near the venter, which has a corded keel. Tubercles mark the points of bifurcation. This genus is found in upper Jurassic strata, worldwide. (K)

Laevaptychus is a bivalved object found in Ammonoid body chambers, or occasionally seeming to close the shell aperture. Their exact function has been much debated but they are now thought to be part of the jaw apparatus. The specimens shown are from Cambridgeshire, UK, and they are found in Jurassic strata, worldwide. (K)

Pavlovia A genus with whorls of circular cross-section and coiling midway between evolute and involute. The ribs are very strong and bifurcate across the venter. A genus from upper Jurassic strata in Europe, Greenland and the USSR. The specimen is 10 cm in diameter. (C)

Titanites is a genus which grows to a considerable size, often over a metre in diameter. The ribs bifurcate across the venter and crowd together in the outer whorl. The coiling is evolute. It is found in upper Jurassic strata in northern Europe, Greenland, Canada and the USSR. (SW)

Titanites

Mantelliceras

Mantelliceras occurs in the upper Cretaceous strata of Europe. It is characterized by moderately involute coiling with coarse ribs developing nodes as they cross the venter. The specimen illustrated is 8 cm in diameter and is from the chalk of northern France; it also occurs in north Africa, India, south-east Asia, Texas (USA) and Brazil. (JM)

Turrilites is unlike many ammonites in the way the shell is helicoid (spiral or screw-shaped). The weak ribs on this specimen develop into strong tubercles on each whorl. The specimen is from upper Cretaceous strata in southern England and is 6 cm long. The genus has a worldwide distribution. (K)

Belemnitella

Passaloteuthis

Actinocamax

Cylindroteuthis

Actinocamax

Sub-class Coleoidea

This group includes squids, cuttles, octopods and the extinct Belemnitida (belemnites). Many groups have an internal shell, others have none. Because of their very hard internal guards belemnites are well known in the fossil record.

Order Belemnitida

These animals were squid-like with tentacles, eyes and an ink sac, and are known from the lower Carboniferous to the Tertiary period. They are characterized by a solid guard, a cigar-shaped internal shell. In most genera this is between 2 and 20 cm long. The phragmacone is less common as a fossil. This is the conical chambered shell, the posterior end of which lies within the end of the guard.

Cylindroteuthis is a large belemnite with a pointed apex and a furrowed cross-section. Near the apex the shell is laterally flattened. The specimen is from the upper Jurassic strata of the Isle of Skye, UK; the genus is from middle and upper Jurassic strata in Europe, North America and Greenland. (K)

Passaloteuthis is medium sized and bullet shaped. The cross-section is round and the guard gradually narrows to a blunt point. The genus is found in lower Jurassic strata, worldwide. (K)

Actinocamax has a medium-sized lanceolate guard on which vascular impressions are commonly found. It also has a squared cross-section. There is a short ventral groove and lines on the surface. It is a genus found in upper Cretaceous strata in Europe, Asia and Greenland. (K)

Belemnitella has a medium-sized, cylindrical guard with lines on the dorsal surface which diverge near the apex. Vascular impressions are often preserved. This genus is from upper Cretaceous strata in Europe, Asia, Greenland and North America. (K)

Acrocoelites is a belemnite commonly found in lower Jurassic sediments in Europe and North America. This bedding plane is covered with fossils which have been orientated by water currents. (L)

Class Gastropoda

The shell, when present in this class, is often coiled in an upward spiral. Some groups have a virtually uncoiled cone-shaped shell, while others have no shell. The body is long with a large head and flattened foot. A characteristic feature of the class is that the body has undergone torsion which twists the nervous system into a figure-of-eight shape, and rotates the mantle cavity and anus in an anti-clockwise manner so that they are higher than the head. The gastropods are the most abundant class of molluscs and are first found in strata of Cambrian age. They are common today and are adapted to live in both marine and fresh water and on land.

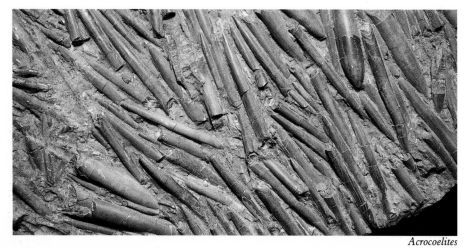

Acrocoelites

Tentaculites

Palaeozoic Gastropods

Maclurites has a maximum length of about 30 cm. The shell has a strongly convex upper surface and a flat base. The umbilicus is deep and steep-walled and growth lines ornament the shell. The specimen is from the middle Ordovician strata of Greenland and the genus is also found in Europe, north-east Asia and North America. (K)

Bellerophon grows to about 10 cm in length. The shell is broad and rounded with a narrow umbilicus. The aperture is flared and there are transverse growth lines. The shell has bilateral symmetry with a ridge around the middle of the whorl. This genus is found in strata of Silurian to Triassic age, worldwide. (K)

Poleumita can have a diameter of up to 4 cm and has a shell with a flat upper surface and numerous thin lamellae and small spines. The specimen is from the Silurian strata in Shropshire, UK, where it is common. The genus is found in Europe and North America. (K)

Straparollus has a variable shape and can either have a high spiral or be much flatter. The shell is much smoother than *Poleumita*. The specimen is from the lower Carboniferous strata of Derbyshire, UK, and the genus is found in strata ranging from Silurian to middle Permian age, worldwide. (K)

Murchisonia has a maximum height of 5 cm. The shell coils in a high spiral with many whorls, but there is little ornament apart from growth lines. This genus is found in rocks of Ordovician to Triassic age in Europe, north-east Asia, south-east Asia, Australia and North America. (K)

Mourlonia is a genus of marine snails which grows to a maximum height of about 4 cm. The whorls are ornamented with spiral cords. It occurs in strata of Ordovician to Permian age, worldwide. (C)

Tentaculites These reasonably common, slender, spiral fossils are possibly of molluscan affinity and have been assigned to the Pteropod group, which contains pelagic gastropods represented in modern oceans by genera only 2 cm long. The individuals on this specimen are 1 cm long and are of Silurian age. They are found in lower Palaeozoic strata, worldwide. (D)

Bellerophon

Murchisonia

Bellerophon

Poleumita

Maclurites

Straparollus

Mourlonia

Symmetrocapulus

Discohelix

Conotomaria

Pseudomelania

Bourguetia

Pleurotomaria

Pseudomelania

Sycostoma

Gyrodes

Cerithium

Gyrodes

Clavilithes

Cornulina

Conotomaria

Mesozoic Gastropods

Pseudomelania has a tall upward spire which is often preserved as an internal mould, as here. The spire is very pointed and there is a slight ridge on each whorl. The genus is from the upper Jurassic strata of Europe. (K)

Pleurotomaria coils in a low spiral with a broad base to the shell. The maximum height is around 12 cm. There is a long slit present on the upper edge from the aperture. The ornament consists of tubercles, spiral bands and growth lines. This genus is found in strata of lower Jurassic to lower Cretaceous age, worldwide. (K)

Conotomaria has a maximum height of some 15 cm. The shell is conical with encircling spiral cords. The whorls are fairly flat, as is the shell base. It is found in rocks ranging in age from middle Jurassic to Palaeocene, worldwide. (K)

Symmetrocapulus is a medium-sized genus with a cap-shaped shell. The ornament consists of radial ribs and concentric folds. It is a genus from Jurassic and Cretaceous strata in Europe. (K)

Discohelix is a small to medium disc-shaped genus. The whorls are sub-quadrate and hardly overlap. The shell is ornamented with spiral threads and tubercles. It has a worldwide distribution and is found in strata ranging from middle Triassic to upper Cretaceous. (K)

Bourguetia is a large, tall genus with a heavy shell. The whorls taper gradually to a blunt point. The shell is smooth apart from growth lines. It is found in Europe in strata of middle and upper Jurassic age. (CP)

Cretaceous and Tertiary Gastropods

Conotomaria (species gigantea) is very large with a height of about 15 cm. There are many whorls in the cone-shaped shell and ornament consists of spiral lines. The genus is found worldwide in middle Jurassic to Palaeocene strata. (K)

Sycostoma is a medium- to large-sized genus with a maximum height of about 7 cm. There is a large body whorl but the others are relatively small. The shell is smooth with ribs on the lower parts. It has a worldwide distribution in strata of upper Cretaceous to Oligocene age. (K)

Cerithium has a small elongated cone-shaped shell with many whorls, and can be up to 3 cm in length. There are growth lines on the shell. This genus is found worldwide in strata from upper Cretaceous to Recent in age. (K)

Clavilithes is a slender genus with a maximum height of about 15 cm. The shell has a long siphonal canal (a narrow slit extending from the aperture) and the whorls have shoulders. The apex is slightly enlarged and the shell is smooth apart from growth lines. It occurs in strata of Palaeocene to Pliocene age in Europe, north Africa and Asia. (K)

Gyrodes has a maximum height of about 5 cm and is a globular shape with an open umbilicus. There are swept-back growth lines and a row of nodes may be present near the join between the whorls. The genus is found worldwide in strata of Cretaceous age. (K)

Cornulina may reach 10 cm in height. The shell is heavily ornamented with spines and tubercles. There is a large body chamber and the other whorls rapidly become smaller towards the apex. The aperture is flared. This genus has a worldwide distribution and is found in strata of Miocene to Recent age. (K)

Galba

Volutocorbis

Potamoides

Murex

Crucibulum

Strombus

Globularia

Turritella

Hippochrenes

Planorbis

Tertiary Gastropods

Murex can have a maximum height of 10 cm, and has a large thick shell with much ornamentation in the form of ribs and spines. The whorls beyond the large body chamber produce a moderately flat shell. The aperture is small compared with the whole shell. This genus has a worldwide distribution and is found in strata of Miocene to Recent age. (K)

Strombus has a maximum height of 15 cm, and has a tough, thick shell. The last whorl is large; the others have an overall conical shape. There are blunt spines around the whorls. This specimen is from the Pleistocene strata of Sicily and the genus ranges from the Eocene to Recent in Europe. (K)

Crucibulum has a medium-sized shell which reaches a maximum height of 5 cm. The shell is cone-shaped and has an open, hollow aperture. Strong ribs radiate from the apex. The specimens are from the Miocene strata of Virginia, USA; the genus ranges from Miocene to Recent in Europe, the West Indies and North America. (K)

Volutocorbis reaches a maximum height of about 5 cm and has a large body chamber with the other whorls forming a tall pointed spire. The whorls are rounded with spiral ridges as ornament. The specimen is from Oligocene strata in Hampshire, UK, and the genus can be found in strata from upper Cretaceous to Recent age in Europe. (K)

Galba reaches a maximum height of about 7 cm and has a tall thin shell of elegant shape. The whorls are convex with very little ornamentation apart from faint growth lines. The genus is from strata of upper Jurassic to Recent age in Europe. (K)

Potamoides can be up to 5 cm high and has a gradually tapering shell with much ornamentation in the form of spines, ribs and growth lines. The specimen is from the Miocene strata of France and the genus ranges from the upper Cretaceous to Recent periods in Europe and Asia. (CP)

Hippochrenes This elegant sea snail has a large extension to the shell producing a wide flat surface above the aperture. The spire is long and pointed with many whorls and the siphonal canal is long. It can grow to a maximum of 15 cm and is a genus from Eocene strata in Europe. (HST)

Turritella A genus characterized by long slender whorls with spiral ribbing. The numerous whorls hardly overlap and the aperture is simple. Growing up to 5 cm in length, this genus ranges from Cretaceous to Recent, worldwide. (C)

Planorbis is characterized by coiling which is almost in a plane spiral, giving a very flat overall shape. The umbilicus is wide and there are very faint growth lines. It grows to a maximum of 3 cm, and occurs in strata of Oligocene to Recent age, worldwide. It is a genus which lives in fresh water and is thus of importance when interpreting the sediments in which it is found. (C)

Globularia This genus grows to about 5 cm and has a large rounded final whorl with a number of smaller whorls forming a fine spire. It occurs in Eocene and Oligocene strata in Europe. (C)

Mactromya

Plagistoma

Plagistoma

Gryphaea arcuata

Plagistoma

Myophorella

Gryphaea dilatata

Lopha

Gryphaea giganteum

Pseudopecten

Hippopodium

Carbonicola

Dunbarella

Anthraconauta

Class Bivalvia

This group of molluscs has two valves which are usually similar and the plane of symmetry passes between them. Well-known bivalves are cockles, oysters, scallops and mussels. There is a superficial resemblance between bivalves and brachiopods, but apart from the fact that both types of shellfish have two shells they have little in common. Their internal organs are different and indeed the brachiopod's two shells differ from one another. The bivalve's two valves are held together by a horny ligament near the beak (umbo) of the shell and there is a hinge mechanism of teeth and corresponding sockets which vary in number and complexity from species to species. The opening and closing of the shell is controlled by two adductor muscles and the ligament. The ligament pulls the valves apart when the muscles relax, and the valves are closed by the pulling muscles. The position and size of the adductor muscles can be determined by studying the muscle scars which are quite obvious oval impressions on the inside of the valves. The line which joins these scars, the pallial line, marks the margin of the attachment of the mantle, which is made of two flaps enclosing the soft body and secreting the calcareous shell. An indentation in the pallial line, the sinus, is present in many species. The presence of a sinus indicates that the animal was a burrower, as the sinus marks the position of retractor muscles required to withdraw the extra large siphons needed by deep burrowing bivalves. Most types have siphons: an in- and exhalent siphon, to draw water with oxygen and food into the shell, and to pass waste out. Bivalves have become adapted to many marine, brackish and freshwater habitats. Many remain anchored to the sea bed; some burrow into the sediment, by using their muscular foot; others live on the sediment and even swim by flapping their valves together. They are usually between 0.5 and 10 cm in width but giants reach 150 cm or more. Bivalve shells are known in the fossil record from Cambrian to Recent times.

Palaeozoic/Mesozoic Bivalves

Gryphaea is illustrated by three different species.
Gryphaea arcuata can be up to about 10 cm long and has an inequivalve shell. The right valve is flat or concave and the left valve is larger and curved into a hooked umbo. A radial groove is on the side of the left valve. Growth lines are very prominent as the main ornament. This species is from the lower Jurassic period and is found worldwide. (CP)
Gryphaea dilatata is wider, larger and more rounded than *G. arcuata*. The shell has thick growth lines. The specimen is from the upper Jurassic strata of Calvados, France. (K)
Gryphaea giganteum has a large rounded shell, with a very concave right valve and convex left valve. The outline is subcircular and there are growth lines on the thick heavy shell. The specimen is from the lower Jurassic strata of Gloucester, UK.
The genus *Gryphaea* is known from the upper Triassic to upper Jurassic strata, worldwide. (K)

Lopha is a small- to medium-sized oyster, which can reach a width of about 10 cm. The shell is thick and heavily ribbed, with convex valves. The valves have a zig-zag margin. This genus is known from Triassic to Recent strata, worldwide. (K)

Hippopodium has a medium to large, thick shell with heavy similar valves. There are thick growth lines and a slight fold near the umbones. This genus is found in the lower Jurassic strata of Europe and east Africa. (K)

Pseudopecten is a large genus with valves up to 20 cm in diameter. It has a subcircular outline with pointed umbones and strong radiating ribs. Growth lines are present and there is one large muscle scar impression on the inside of each valve. Pectens (Scallops) are one of the few groups of bivalves which can swim by flapping their valves together. This genus is found in Jurassic strata in Europe, South America and the East Indies. (K)

Plagistoma has a maximum diameter of about 12 cm, with a smooth shell ornamented by growth lines. There is a straight edge on one margin and a curved opposite side and rounded posterior. A small wing is near the umbones on the anterior side. The genus is from middle Triassic to Cretaceous strata, worldwide. (K)

Myophorella can be up to 10 cm wide and has a subtriangular shell outline with umbones pointing inwards. The valves are thick and beautifully ornamented with rows of heavy tubercles. The genus can be found in strata of lower Jurassic to lower Cretaceous age, worldwide. (K)

Mactromya is a medium-sized genus with an oval outline. There are numerous growth lines and the umbones are strongly curved towards and against each other. There is a pronounced posterior gape in this burrowing bivalve. The genus is found in lower Jurassic sediments in Europe and north Africa. (K)

Dunbarella is a genus of marine bivalves, often found in the same upper Carboniferous strata as goniatites. It is a thin-shelled genus with delicate ribs and growth lines. The overall shape is similar to that of Pectens. It is found in Europe and North America. (S)

Carbonicola is a small genus, which grows up to 6 cm long. The umbo is directed forwards and there are growth lines following the shape of the valves. Living in non-marine conditions, this genus is similar to modern genera like *Unio*. *Carbonicola* was not attached to the substrate and probably moved through the soft mud. It is found in upper Carboniferous strata in Europe and the USSR. (S)

Anthraconauta Valves of this genus crowd a bedding plane from the upper Carboniferous. This non-marine genus is used, along with several others, for subdividing the coal bearing strata for correlation. The range of this genus is upper Carboniferous to Permian and it is found in western Europe and the USSR. (S)

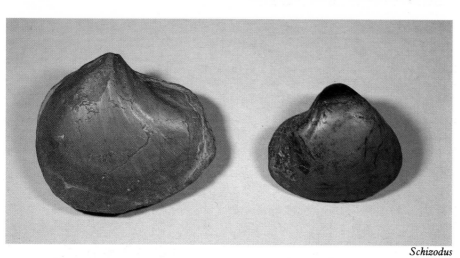

Schizodus

Schizodus is a small- to medium-sized genus which grows to about 5 cm. The shell is smooth, being ovoid with posterior elongation. It ranges through the Carboniferous and Permian periods, and has a worldwide distribution. (K)

Mesozoic Bivalves

Gervillella can be up to 20 cm long and has a slender shell which is elongated posteriorly. The umbones are sharp and pointed with a short posterior wing. Fibrous ligament pits and concentric growth lines are present. This genus ranges from Triassic to Cretaceous age, worldwide. (K)

Eoradiolites has a conical to tube-shaped shell with the right valve attached. There is ribbed ornamentation. The diameter of this bizarre bivalve shell can be up to 5 cm. The specimen is from Cretaceous strata in Iowa, USA, and the genus is also found in Europe, Africa and Asia. (K)

Hippurites is another tube-shaped genus which grows up to 12 cm in diameter. The right valve is attached and the left valve acts as an operculum (lid). Superficially this genus resembles a coral. The specimen is from northern France and the genus ranges through Europe and north Africa to south-east Asia, in upper Cretaceous strata. (K)

Inoceramus grows to about 12 cm in length and has a convex shell with large concentric ridges and growth lines. The umbones often have large wings. This genus is found in Jurassic and Cretaceous strata, worldwide. (K)

Spondylus may be up to 12 cm long and has a straight hinge line with neat, pointed umbones. There are many striations near the umbones and strong ridges run from the beak to the shell margin. These are often spinose. The genus has a worldwide distribution, and occurs in strata of Jurassic to Recent age. (JHF, HST)

Modiolus can be up to 22 cm long and is not unlike the well-known mussel in outline, though it is somewhat more elongated in some species. There is a long ligament and many growth lines. This genus occurs in strata from Devonian to Recent in age, worldwide. (K)

Lopha is characterized by many strong ribs and, in this species, by a relatively narrow, elongated shell. One valve is convex, the other concave. The specimen is from the Cretaceous strata of northern France and the genus is found worldwide in rocks of Triassic to Recent age. (K)

Lima has a sub-trigonal shape with a short hinge margin and slightly convex valves. The ribs have small sharp nodes and there are obvious growth lines. Near the umbo there are slight wings. This genus occurs worldwide in strata of Jurassic to Recent age. (K)

Pholadomya reaches a maximum length of about 12 cm. The valves have an elongate oval outline. There are prominent umbones and the valves are convex with a posterior gape. The pallial line has a deep sinus, typical of a burrowing genus. It is found in strata of upper Triassic to Recent age, worldwide. (K)

Nuculana is a small genus reaching only some 2 cm in length. The outline is sub-trigonal with a backward pointing beak. Growth lines are the

Nuculana

Spondylus

Pholadomya

Modiolus

Spondylus

Modiolus

Inoceramus

Hippurites

Gervillella

Lima

Eoradiolites

Lopha

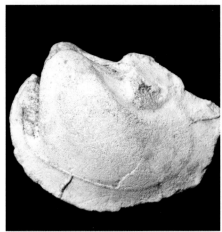

Laevitrigonia

only ornaments. It is a common genus in lower Jurassic strata, but can also be found worldwide in strata from Triassic to Recent age. (K)

Tertiary Bivalves

Venericardia is characterized by a thick, heavy shell up to 15 cm wide. Both valves are convex with well-formed growth lines and radiating ribs. One of these specimens shows muscle scars and heavy dentition with two curved teeth. The shell margin is crenulated. This genus is common in strata of Palaeocene to Eocene age in Europe, Africa and North America. (K)

Arctica is up to 10 cm wide and has a subcircular outline with prominent curved umbones. The shell margin is smooth and there are concentric growth lines. The specimen shows three large central teeth and long lateral teeth. There is no pallial sinus on the line joining the adductor muscle scars. Found in North America and Europe, this genus ranges from Cretaceous to Recent age. (K)

Teredo has an extremely small shell and burrows into wood. One of the two specimens is a mass of burrows from the Eocene London clay, the other is of two calcified burrows from the Miocene strata of Virginia, USA. The circular burrows often have shell fossils at one end. With a worldwide distribution, this genus is of Eocene to Recent age. (K)

Corbicula is a small to medium-sized shell, up to about 3 cm wide. The outline is almost oval and the umbones curve anteriorly. There are fine ribs on an otherwise smooth shell. It is a worldwide genus from lower Cretaceous to Recent strata. (K)

Laevitrigonia This specimen is preserved as an internal cast which shows clearly the muscle scars and the pallial line linking them. The valve is very inequilateral and has a pronounced umbo and posterior elongation. The specimen is 5 cm across and comes from the upper Jurassic strata of southern England. The genus is found in strata of this age in Europe and the Middle East, Asia and east Africa. (C)

Protocardia These small bivalves have been weathered to produce a heart-shaped cross-section. This genus grows to 8cm wide, and occurs in upper Triassic to upper Cretaceous strata in Europe, Africa and North and South America.

Corbicula

Teredo

Venericardia

Arctica

Teredo

Protocardia

Brachiopods

Platystrophia

Rafesquina

Glyptorthis

Plaesiomys

Rafesquina

Orthis

Lingulella

Phylum Brachiopoda

These shellfish are common as fossils; during the Palaeozoic and Mesozoic eras they made up a very large part of the sea-bed fauna. They are not as numerous today, however, as they have been at many times in the past.

They first appear in great numbers in the fossil record in the Cambrian period and quickly develop in a great variety of forms adapted to different sea-bed habitats. Some burrowed into the soft mud, others were fixed to rocks. On average, brachiopods are about 0.5 cm to 8 cm long, but some species reach 37.5 cm.

Superficially, brachiopods resemble bivalved molluscs. The main similarity is that both groups have two valves. In brachiopods, however, the two valves are dissimilar. The pedicle valve is generally larger than the brachial valve and has a hole in the posterior end through which a fleshy stalk, the pedicle, protrudes to anchor the animal to the sea bed. The smaller brachial valve contains an organ unique to this phylum, the lophophore. This bilaterally symmetrical, curved organ is covered with cilia and as it vibrates so water is moved in and out of the open anterior end of the shell. Food matter in the incoming water is caught on the lophophore and transferred to the mouth. A fleshy mantle lining the inside of the valves contains the soft organs and secretes new shell. The details of the soft anatomy are known from modern living brachiopods, but many fossils provide details of the inside of the shell indicating how and where muscles were attached. In some species the brachidium, an internal calcareous framework which supported the lophophore, is preserved. Two classes of brachiopods are recognized: the Inarticulata, which have musculature but no hinge mechanism, so they are unable to open and close their shell, and the Articulata, which have a complex system of muscles and hinge mechanism that enables them to move one valve relative to the other.

Brachiopods are very common as fossils in the lower Palaeozoic strata and they account for over 75 per cent of the fossilized Silurian sea-bed organisms. These organisms are easily preserved in limestones and shales, especially those which were formed on the extensive continental shelves. In this example, from the Silurian strata of Shropshire, UK, a specimen of the brachiopod *Leptaena* is on the left side of the picture, while smaller specimens of *Orthis* are to the right of the centre. Along with the brachiopods there are fossils of crinoids, corals and bryozoans. (L)

Cambrian and Ordovician Brachiopods

Lingulella can be 1 to 3 cm long as an adult and is an inarticulate brachiopod. The shell is almost oval in outline, with many fine growth lines. The two valves are often disarticulated before fossilization. *Lingulella* burrowed into the sea bed and was anchored to the floor of the tubular burrow by a long pedicle. The specimens shown are from Portmadog, Wales (UK) and are of Cambrian age. The genus ranges from Cambrian to Ordovician and is found worldwide. (K)

Rafesquina is about 3 to 4 cm wide when adult. The shell has a convex pedicle valve and a concave brachial valve, the whole outline being semi-circular. There are radiating ribs on the valves and a small beak on the pedicle valve. This specimen is from the upper Ordovician strata of Cincinnati, Ohio (USA), and the genus has a worldwide distribution in Ordovician strata. (K)

Plaesiomys reaches a width of about 3 cm when mature. The pedicle valve is convex, the brachial valve concave. The overall shape of the shell is semi-circular and thin ribs radiate from the umbo. The specimens are from Iowa, USA, and the genus is found in Ordovician rocks worldwide. (K)

Platystrophia grows to a width of about 5 cm. It has an almost rectangular outline with a furrow in the pedicle valve and a fold in the brachial valve. There are obvious sharp ribs radiating from the umbo. The specimens illustrated are from the Ordovician strata of Indiana, USA, the genus ranging from the Ordovician to the Silurian with a worldwide distribution. (K)

Glyptorthis has a width of about 2 cm when adult and is almost rectangular in outline. There is much ornamentation on the shell with growth lines and ribbing. The genus is found worldwide in Ordovician and Silurian rocks. The specimens illustrated here are from Ohio, USA. (K)

Orthis has a width of 2 cm when mature and its shell is almost circular in outline. The hinge line is short and the pedicle valve is convex while the brachial valve is convex or flat. Radiating from the umbo are stout ribs with sharp crests. Found in Ordovician rocks, *Orthis* has a worldwide distribution. (K)

Silurian and Devonian Brachiopods

Shaleria has a maximum width of about 2.5 cm along the straight hinge line. The margin of the shell flares outwards at the hinge and there are numerous radiating ribs across the shell. The specimens are from Gotland, Sweden; the genus is of Silurian age and is found in the northern hemisphere. (K)

Protochonetes reaches a maximum width of about 2 cm at the hinge line, which is straight. The shell is a broad semi-circular shape with numerous small fine ribs. This specimen is from Shropshire, UK, and the genus is found worldwide in strata of Silurian age. (K)

Salopina is about 0.5 cm wide with a roundish shell outline and a pointed umbo, and with ornamentation of fine ribs and growth lines. This genus is from the Silurian rocks of Europe. Both *Salopina* and *Protochonetes* are on the same specimen. (K)

Leptaena can be from 2 to 5 cm wide with a convex pedicle valve and a flat brachial valve. Both valves curve obviously near the anterior margin of the shell. Ornamentation consists of fine ribs and concentric wavy ridges. The specimen is from Silurian rocks at Hereford, UK; the genus ranges from the Ordovician to Devonian and is found worldwide. (K)

Eospirifer

Mucrospirifer

Strophonella

Leptaena

Gypidula

Shaleria

Strophonella

Protochonetes and *Salopina*

Strophonella grows to a shell width of between 3 and 5 cm. The hinge line is straight and the overall outline somewhat shield-shaped. There are radial ribs. The genus is of Silurian to Devonian age and is found worldwide. The specimens are from the Devonian strata of Iowa, USA. (K)

Gypidula has a roughly circular outline and reaches about 4 cm in width. The umbo on the pedicle valve is very pronounced and there are some ribs on the ventral surface. The genus is of Silurian to Devonian age and has a worldwide distribution. (K)

Eospirifer grows to a maximum width of about 12 cm and has a very wide hinge line. There is a small pedicle opening and a fold in the brachial valve with a corresponding groove in the pedicle valve. The beak in the pedicle valve is very obvious and ornamentation consists of many rounded ribs and growth lines. The specimens are from Devonian strata in Canada; the genus has a worldwide distribution. (K)

Mucrospirifer attains a maximum width of about 6 cm. It has a very wide, almost pointed hinge line and a small pedicle opening. The brachial valve has a fold and the pedicle valve a groove, both being crossed by many rounded ribs and wavy growth lines. The specimens are from the Devonian rocks of Hamilton, USA, and the genus is found worldwide in Devonian strata. (K)

Oolitic limestone quarry

Pugilis is medium to large in size, reaching about 8 cm width. The hinge line is straight and the shell has growth lines and ribs. The pedicle valve is convex, folding round to the umbo, and may be spinose. The genus is found in Carboniferous strata in the UK. (K)

Imbrexia reaches a maximum width of about 8 cm along the hinge line. The beak projects forward of the hinge line and the shell is covered with numerous ribs. The genus is from lower Carboniferous strata and has a worldwide distribution. (K)

Spirifer reaches about 12 cm width when adult. The hinge line is wide and the pedicle opening is small. The pedicle valve has a groove and the brachial valve a fold. The beak in the pedicle valve is distinct and there are many ribs and growth lines. The specimens are from the Carboniferous strata of Derbyshire, UK; the genus has a worldwide distribution. (K)

Pustula has a width of about 12 cm when mature, and is similar to *Productus* in many features. The hinge line is straight and there are growth lines and faint ribs on the shell. The pedicle valve is convex, the brachial valve flat. The specimen is from Staffordshire, UK, and the genus is found in Carboniferous rocks in Europe. (K)

Davisiella grows to about 12 cm width, and has a robust shell with a convex pedicle valve and flat brachial valve. Faint ribs run across the shell but growth lines are rarely visible. The specimen is from North Wales and the genus is found in Carboniferous strata in Europe. (K)

Schizophoria has a maximum width of 5 cm when mature. The shape is biconvex with a furrow in the pedicle valve and a fold in the brachial valve. There are numerous small ribs. The specimen is from the Carboniferous strata of France; the genus is found worldwide in Permian and Carboniferous strata. (K)

Unispirifer has a medium to large shell reaching a width of 12 cm across the straight hinge line.

Carboniferous Brachiopods

Productus grows to a width of about 4 cm. It has a very convex pedicle valve and a concave or flat brachial valve. There are simple ribs, wavy growth lines and a short, straight hinge line. The specimen is from Derbyshire, UK, and the genus is found in Europe and Asia in strata of Carboniferous age. (K)

Lingula is an inarticulate brachiopod which grows to a length of about 3 cm. It has an equivalve shell with an almost oval outline and very fine ribs and growth lines. It is a burrowing genus which in life has a pedicle as long as the shell to anchor it in the burrow. The specimens are from Northumberland, UK, and this is one of the longest ranging genera, being found in rocks of Ordovician to Recent age worldwide. (K)

Pugnax has a width of 3 cm when fully grown. The outline of the shell is triangular, and there is a deep furrow in the pedicle valve and a large fold in the brachial valve. Very faint ribs cross the shell. The genus ranges from Devonian to Carboniferous age and is found throughout Europe. (K)

Rhynchonotreta

Obovothyris

Epithyris

Sphaeroidothyris

Waltonithyris

Rugitela

Pseudoglossothyris

Morrisithyris

Ornithella

The shell is biconvex and has many ribs, some of which split in two (bifurcate). The genus has a worldwide distribution in strata of lower Carboniferous age. (K)

Brachythyris reaches a maximum width of about 7 cm. The hinge line is less wide than the maximum width and there is an obvious beak on the convex pedicle valve. The brachial valve is also convex and the margin of the shell is distinctly folded. The genus is found in Carboniferous strata worldwide. (K)

Rhynchonotreta is a genus of small triangular shaped brachiopods from the Silurian strata of Scandinavia, Britain, Czechoslovakia and North America. The shell is ornamented with strong ribs which radiate from the umbo and the hinge line is strongly folded. (C)

Oolitic limestone quarry   Fossil localities are often made-made. This quarry in Yorkshire, UK, is rich in brachiopods and molluscs.

Jurassic Brachiopods

Epithyris grows to a length of about 3 cm and has an obvious beak with a large pedicle opening. There are two folds in the brachial valve and a groove in the pedicle valve. The ornamentation consists of growth lines; there are no ribs. The specimen is from the Cotswold Hills, UK, and the genus is found in the middle Jurassic strata of Europe. (K)

Pseudoglossothyris is a large brachiopod which grows to a length of up to 10 cm. The pedicle valve has a definite umbo with a large pedicle opening. The shell is biconvex with a generally oval outline and has concentric growth lines. The genus is from the middle Jurassic strata of Europe. (K)

Obovothyris has a width of 1.5 cm when adult. The shell is biconvex and the umbo curves towards the brachial valve. The pedicle opening is small. There are no ribs and the growth lines are almost pentagonal in outline. The genus is from the middle Jurassic strata of Europe. (K)

Sphaeroidothyris is a small rounded brachiopod with a diameter of about 2 cm when mature. The shell is biconvex and has faint

growth lines. The genus is from Jurassic strata in France, Germany and the UK. (K)

Waltonithyris is about 3 cm long when mature. There is an obvious umbo on the pedicle valve, which has two furrows separated by a ridge running from the anterior margin. From the Jurassic strata of Europe. (K)

Morrisithyris is an elongate genus which may reach a length of 6 cm. The posterior end of the shell has a triangular shape while the anterior part is folded with a deep furrow on the brachial valve. From the Jurassic strata of Europe. (K)

Rugitela reaches a length of about 4 cm when adult. The shell has an elongate pentagonal outline and is biconvex. The beak is slightly curved and there is a small pedicle opening. Growth lines cover the shell. From the Jurassic strata of Europe. (K)

Ornithella has a maximum length of 3.5 cm. It has a pentagonal outline, the anterior margin being straight. The beak curves and there is a small pedicle opening. The shell is smooth. The genus is from the Jurassic strata of Europe. (K)

Eudesia

Dictyothyris

Isjuminella

Spiriferina

Torquirhynchia

Tetrarhynchia

Acanthothyris

Goniorhynchia

Spiriferina

Plectothyris

Jurassic Brachiopods

Plectothyris reaches a width of 3 to 4 cm. The shell is biconvex with an almost circular outline. There is an obvious beak and pedicle opening and the anterior margin of the shell has rounded ribs. The genus is found in Jurassic rocks in the UK. (K)

Dictyothyris grows to a width of 1.5 cm when adult. The shell outline is almost pentagonal and there are two large folds on the pedicle valve. The shell is ornamented with small ribs and growth lines. The genus is from the Jurassic strata of Europe. (K)

Acanthothyris has a maximum width of about 1.5 cm and its outline is almost circular. There is a small sharp beak and many ribs which may have short spines. The specimen is from Somerset, UK; the genus is found in Europe and Asia in strata of Jurassic age. (K)

Tetrarhynchia has a maximum width of 1.5 cm when adult. The shell has a sub-triangular outline; the pedicle valve is convex with a small pointed beak. The ribs become very pronounced and the shell margin is a strong zig-zag. Near the umbo the shell is much smoother. The genus is found in Jurassic strata in Europe and North America. (K)

Goniorhynchia can be 3 cm wide when mature; the shell is biconvex and covered with strong sharp ribs. The beak is definite and there is a small pedicle opening. The genus is found in Jurassic strata in Europe. (K)

Torquirhynchia reaches a maximum width of 4 cm when adult and is biconvex, with an almost rounded outline. There is a strong beak and small pedicle opening, and coarse simple ribs cross the shell. The genus occurs in strata of upper Jurassic age in Europe, west of the Alps, and Russia. (K)

Spiriferina has a width of 2 to 3 cm when mature. The shell is biconvex with large ribs. There is a fold in the brachial valve and a furrow in the pedicle valve. Some individuals have small spines. The specimen is from Somerset, UK; the genus is from the Triassic and Jurassic periods and has a worldwide distribution. (K)

Eudesia has a width of 2 cm when adult. The shell outline is almost oval and there is a strong beak with a large pedicle opening. The ribs are large and sharp. The specimen is from Normandy, France, and the genus is found in Jurassic strata in Europe, Asia and North America. (K)

Stenoscisma

Stiphrothyris

Orbirhynchia

Kingena

Terebratella

Pygope

Cyclothyris

Concinnithyris

Isjuminella is a large thick-shelled brachiopod which may be up to 10 cm wide. It has a globose shell covered with many stout ribs, both valves being convex. The genus is found in Jurassic strata in Europe and the USSR. (K)

Stenoscisma is a genus of small brachiopods which reaches about 2 cm in diameter. It has a pointed umbo and there are strong ribs across the valves. These specimens are internal casts from the reef limestones of Permian age in Durham, UK. The genus is found in strata from middle Devonian to upper Permian age in Europe. (C)

Stiphrothyris is a genus from the middle Jurassic strata of Europe. The specimen illustrated shows certain features of the phylum. The pedicle opening (or foramen) is clearly seen at the top of the picture. The specimen has been opened so that the loop can be seen. This is a calcareous support for the lophophore, a feeding structure which has arms with ciliated tentacles, causing water currents to pass into the shell. Microorganisms are filtered from the water and passed along the lophophore to the digestive system. (K)

Jurassic and Cretaceous Brachiopods

Concinnithyris has a width of about 4 cm when adult. The shell is smooth and almost circular in outline. Both valves are convex and there is a large pedicle opening. The genus is found in Cretaceous strata in Europe, North America and Asia. (K)

Cyclothyris grows to a width of 3 cm. The shell is roughly triangular in outline and there is an obvious beak and pedicle opening. The shell is biconvex and covered with many ribs and growth lines. The specimen is from Devon, UK; the genus is found in Cretaceous rocks in Europe and North America. (K)

Orbirhynchia is about 1.5 cm wide when mature. The shell has a roughly circular outline

and is globular. There are ribs and a small beak with pedicle opening. The genus is found in Cretaceous rocks in Europe. (K)

Terebratella reaches a length of about 4 cm when adult. The shell is oval in outline and is biconvex. There is a small pedicle opening below the beak. The shell has many ribs and some growth lines are obvious. The genus is from Jurassic to Recent strata with a worldwide distribution; the specimen is from strata of Cretaceous age in France. (K)

Kingena has a maximum length of 2 to 3 cm and is roughly circular in outline. There is a large

pedicle opening and the shell is smooth with growth lines but no ribs. The specimen is from rocks of Cretaceous age in Texas, USA, and the genus has a worldwide distribution in Cretaceous strata. (K)

Pygope is a large, strangely shaped brachiopod. It has a triangular outline and each side may be up to 8 cm long. The shell margins are almost vertical and there is a deep groove running through the centre of the shell which develops into a perforation. In some forms this leads to a bi-lobate shell shape. The specimens are from Verona, Italy, and the genus is found in Jurassic and Cretaceous strata in Europe. (K)

Hemicidaris

Pygurus

Pseudodiadema

Holectypus

Nucleolites

Acrosalenia

Pygaster

Stomechinus

Balanocidaris

Pedina

Clypeus

Phylum Echinodermata

Of the eight classes within this phylum only four, the Echinoids, Crinoids, Asteroids and Ophiuroids are common in the fossil record. The phylum includes such creatures as the well-known starfish and sea-urchin, and all the organisms in the phylum are marine. Most of these creatures have a calcareous endo-skeleton with five-fold (pentameral) symmetry. The phylum is also characterized by a water vascular system – an internal combination of pipes filled with fluid which is linked to the external tube feet. These are used for movement, feeding and breathing.

The class Echinoidea (sea-urchins), has a test (or shell) which varies from being circular in plan to being oval or heart-shaped. This is made up of interlocking calcareous plates which join along zig-zag sutures. When alive the test is covered with spines. These can vary from being many and hedgehog-like, to being few and club-shaped. The plates which form the test are arranged in bands which encircle the whole structure in many genera, but in some forms these bands are atrophied. The bands of narrower plates (the ambulacra) alternate with broader bands (the interambulacra) and both bands run from the small circular mouth region (the peristome) on the underside of the test to the anal region (the periproct) on the upper surface. In some of the irregular forms the positions of the mouth and anus are not central. The ambulacral plates are porous with small holes through which the tube feet pass, but the interambulacral plates are non-porous, having tubercles on to which spines may join with a ball-and-socket joint. This basic shell structure has a pentameral symmetry in the group called the regular echinoids, but in the irregular forms it is adapted in a number of ways. The circular outline changes to become, for example, oval or heart-shaped; the central position of the mouth and anus changes and the ambulacra become shorter and petaloid. The test now has a bilateral symmetry. Such irregular echinoids are adapted to specialized habitats, such as burrows, whereas the rounded regular forms live on the sea bed. The very flattened 'sand-dollars' have a shell adapted to living on the sea bed with currents of sediment passing across and over the test. Echinoids are known in the fossil record from the Ordovician to Recent periods and are common members of the marine fauna today.

The Crinoidea (sea-lilies) are common fossils in the Palaeozoic and Mesozoic eras, and their broken remains build great masses of limestone. Because of their delicate calcareous structure, crinoids are rarely found entirely preserved. The soft crinoid body is held in a limy calyx built of basal and radial plates and supported a short distance above the sea bed on a stem made of calcite ossicles. This stem has roots which anchor the organism to the sea bed. Above the calyx are arms and an anal tube. Food can be directed by cirri on the smaller arms down food-grooves to the mouth. Some crinoids are free-swimming, others cling to seaweed. The class first appears in lower Cambrian strata and they still live in modern oceans.

The class Asteroidea is uncommon as fossils and is often preserved in a fragmented state. This class of starfish ranges from Ordovician to Recent age. They have five arms (or multiples thereof), which can be short and stumpy or long and tapering. These radiate from a central disc, and the undersides of the arms have numerous tube-feet. The central body is broadly similar to that of the echinoids.

The Ophiuroidea contain delicate 'brittle stars'. The skeleton is formed of ossicles, which in the central disc are plate-like, and rounded in the five flexible arms. Tube-feet are not present. Though they are not common as fossils, their broken fragments are occasionally abundant in strata ranging from Ordovician to Recent age.

Jurassic and Cretaceous Echinoids

Nucleolites is an irregular genus with a rounded to slightly heart-shaped test, and of small to medium size. The ambulacra are petaloid and there are numerous small tubercles on the interambulacra. The symmetry is bilateral, and the peristome is anterior. The genus is found in strata ranging from middle Jurassic to upper Cretaceous in Europe. (K)

Acrosalenia is rounded in outline and has very large tubercles on the interambulacral areas. The narrow ambulacra have smaller tubercles. The central position of the peristome and periproct allow this genus to be placed in the regular group. Ranging from the Jurassic to the Cretaceous age, *Acrosalenia* is found in Africa and Europe. (K)

Hemicidaris has a sub-spherical test with radial pentameral symmetry and a central periproct, the plates surrounding the anus being slightly raised. There are larger tubercles on the oral surface. This genus is of lower Jurassic to upper Cretaceous age, and is found in Asia, Africa, Europe and North America. (K)

Pedina is a regular echinoid of sub-spherical outline, with typical pentameral symmetry radiating from the central mouth and anus. The test is flattened and covered with small tubercles. *Pedina* is found in strata of Jurassic to Miocene age in Europe, Africa and South America. (K)

Holectypus has a circular outline with a hemispherical side view. The oral surface is concave or flat and the periproct is marginal in this irregular genus. There are larger tubercles

Lapworthura

on the oral surface and the ambulacra are non-petaloid. This genus is found in strata of lower Jurassic to upper Cretaceous age, in Europe, North America, Cuba, Iran, north Africa, Japan and Venezuela. (K)

Clypeus is a medium to large genus with a very flattened test. There are petaloid ambulacra and broad interambulacra with rows of minute tubercles. This genus is from strata of Jurassic age in Europe, Africa and Australia. (K)

Pygaster is a medium to large genus with an oval region on the upper (aboral) surface which contains the anus. The test has a roughly pentagonal outline and is covered with small, flat tubercles. The interambulacra are broad. *Pygaster* is a genus from the Jurassic and Cretaceous strata of Europe. (K)

Pygurus has a pentagonal outline and a test which is somewhat flattened. The petaloid ambulacra have elongated pores on their outer margins. The mouth is central and a ridge runs from it to the posteriorly positioned anus. This irregular genus has a worldwide distribution in strata of Cretaceous to Eocene age. (K)

Pseudodiadema is a characteristically regular echinoid with a central mouth and anus and entire ambulacra. There are very stout tubercles for spines. It is a genus which is found in strata of Jurassic and Cretaceous age in Europe, north Africa, India, the USA and Brazil. (K)

Stomechinus is a regular genus of small to medium size. The narrow ambulacra have two rows of tubercles. The interambulacra are wide with well-spaced tubercles. The central mouth is in a large peristome. From the Jurassic strata of Europe, north Africa, the USA and Brazil. (K)

Balanocidaris Two spines are illustrated from this genus. They show the characteristic narrow base and swollen bulbous structure with longitudinal ridges. A genus from the Jurassic strata of Europe, Egypt, north Africa, Japan, and California (USA). (K)

Lapworthura is an ophiuroid with a similar structure to Recent genera but it occurs in strata of Ordovician and Silurian age. This specimen is from northern England; the genus is also found in Scotland and Australia. (K)

Micraster cortestudinarium

Micraster decipiens

Micraster gibbus

Echinocorys

Holaster

Cardiaster

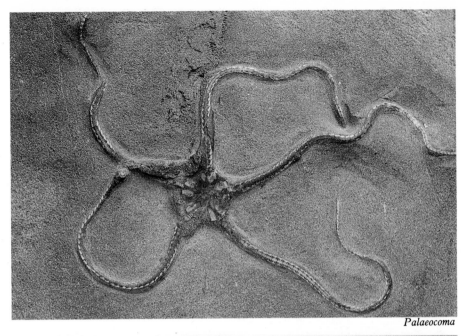

Palaeocoma

Palaeocoma is a well-known genus of ophiuroid (brittle-stars), with slender arms and a relatively small central disc, which grows up to 2 cm in diameter. The arms are made up of diamond-shaped plates. It is very similar to living brittle-stars and is a locally common fossil in Jurassic and Cretaceous strata in Europe. (L)

Cretaceous and Tertiary Echinoids

Holaster is an irregular genus with a slightly heart-shaped outline. The anus is near the point of the test, the peristome is anterior. There are large tubercles on the under surface while those on the upper surface are very small. The pores are elongate and slit-like. The broad interambulacra alternate with sub-petaloid ambulacra. It has a worldwide distribution and is found in strata of lower Cretaceous to Eocene age. (K)

Echinocorys is an irregular genus with a moderately large, high test which has a flat base. The mouth is in an anterior position and the anus is on the underside at the posterior end. There is a slight ridge running from the anus to the mouth, and the ambulacra, which are not petaloid, have rounded pores. There are very small tubercles on the broad interambulacra. This genus is found in upper Cretaceous strata in Europe, Asia Minor, Madagascar, North America, Cuba and Georgia (USSR). (CP)

Cardiaster is a medium to large, irregular, heart-shaped genus with ambulacra entire and non-petaloid. Both the mouth and anus are non-central. The genus is from the Cretaceous strata of Europe. (K)

Micraster is a genus of irregular echinoids which are common in strata of Cretaceous age. The test is medium-sized and heart-shaped with petalloid ambulacra which have rounded pores. There is a deep anterior notch or groove with the mouth just below it. The anus is posterior, relatively high on the test. The roughly oval to triangular area between the mouth and anus on the oral surface is called the plastron; this is a development of the posterior ambulacrum. The genus *Micraster* has a well-known evolutionary sequence, the species of which are used to zone the middle and upper Chalk of Europe. Three species are shown: *M. cortestudinarium* (preserved in flint), *M. decipiens* and *M. gibbus*. The position of the anus is easily seen in the specimen of *M. decipiens*. The genus ranges from the upper Cretaceous to Palaeocene periods, with a worldwide distribution. (HST, K)

Dendraster is a genus of irregular echinoids with an ovoid outline and petaloid ambulacra. The periproct is near the margin of the test. The food grooves bifurcate. This genus is found in strata of Pliocene to Recent age in the USA. (C)

Lovenia has a heart-shaped shell with petaloid ambulacra on the posterior region and a non-petaloid anterior ambulacrum. There are recessed tubercles on the interambulacra and the peristome is anterior. An Eocene to Recent genus from Australia. (C)

Dendraster

Lovenia

Camerogalerus

Parmulechinus

Conulus

Stigmatopygus

Cidaris

Hemipneustes

Schizaster

Phymosoma

Stereocidaris

Clypeaster

Seaside chalk cliffs on the island of Corfu

Cretaceous and Tertiary Echinoids and Asteroid

Camerogalerus has a rounded outline with a flat oral surface and a high cone-shaped apex. The ambulacra are narrow and entire, the interambulacra broad. *Camerogalerus* is a genus from Cretaceous strata in Europe. (K)

Conulus is an irregular genus with a dome-shaped aboral surface and a flat oral surface. There are non-petaloid ambulacra, and many small tubercles are present on the interambulacra. The mouth is central but the anus is on the margin of the test. This genus is found in the upper Cretaceous strata of Europe, Africa, North America and Asia. (K)

Stigmatopygus is a small genus with a circular outline and a conical apex, from the Cretaceous strata of India, Europe and Africa. (K)

Parmulechinus has a strange flat, discoidal shape which is virtually two-dimensional. The mouth is central. From the Oligocene strata of Europe and north Africa. (K)

Cidaris is a regular, roughly spherical, medium-sized echinoid with characteristic radial pentameral symmetry. The ambulacra are narrow with simple plates containing single pairs of pores. Each of the interambulacral plates has a large tubercle. The spines are long and ornamented. The mouth and anus are central. This genus is found in strata ranging from upper Triassic to Recent age and has a worldwide distribution. (K)

Schizaster is an irregular heart-shaped genus with petaloid ambulacra and non-central mouth and anus. The periproct is below the pointed end of the test (posterior). The mouth is below a

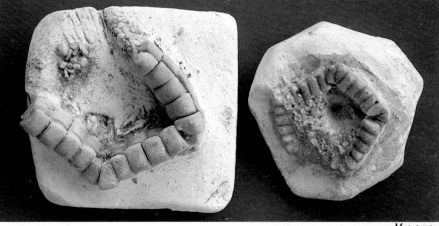

Metopaster

furrow in the anterior margin. This genus is found worldwide in strata of Eocene to Recent age. (K)

Hemipneustes is a large irregular genus with an ovoid test and high convex upper surface. The anterior ambulacrum is in a groove and the peristome is anterior; the periproct is marginal. This genus is found in the Cretaceous strata of northern Europe. (K)

Phymosoma is a regular genus with a rounded test, and has a central peristome and periproct. The ambulacra are entire and the test bears large tubercles, to carry stout spines. From the Jurassic and Cretaceous strata of Europe. (K)

Stereocidaris is a small- to medium-sized regular genus with a globular test of circular outline. Long heavy spines are carried by the tuberculate interambulacra. These spines are rarely present on fossil material, though they are often found separately. It is found in Cretaceous

to Recent strata throughout Eurasia; today it lives in the Indo–Pacific region. (K)

Clypeaster is a very large irregular genus which has an oval test with five lobes around the margin. The concave oral surface has broad petaloid ambulacra and a central peristome. The anus is at the posterior end of the oral surface. The mouth is in a deep hollow. It is found worldwide in strata of Upper Eocene to Recent age. (K)

Metopaster is a genus of asteroid (starfish) which lacked arms and consisted of an outer pentagonal ring of large plates and many smaller plates forming the ventral and dorsal surfaces. The maximum diameter is about 6 cm. It is found in Europe in strata of Cretaceous to Miocene age. (JHF, K)

Sea-side chalk cliffs on the island of Corfu contain fossils of a variety of creatures including echinoids and molluscs.

143

Platycrinites

Pentacrinites

Dimerocrinites

Periechocrinus

Encrinus

Platycrinites

Woodocrinus

Crinoidea

Pentacrinites is a well-known Mesozoic genus which has a small calyx made up of two series of plates. There can be many long arms with pinnules and cirri occur on the arms and the stem. The ossicles are star-shaped. The stem may be over a metre long. It is often preserved only as masses of ossicles. Modern forms of this genus break free from the sea bed and swim. Found in Europe and North America, it occurs in strata of Triassic to Recent age. (K, L)

Encrinus has a rather squat calyx but still shows good pentameral symmetry. The long arms are robust at the base and have pinnules. The stem is circular in section. This is a well-known genus in the German Triassic strata, and ranges through Europe (though not the UK) in strata of middle and upper Triassic age. (K)

Dimerocrinites has a cone-shaped calyx and the arms are branched with pinnules. The long slender stem is made of rounded ossicles. The calyx may be 3 cm in diameter. It is found in strata of middle Silurian to middle Devonian age. (K)

Platycrinites is characterized by a flat, twisted stem and a cup-like calyx. The base of the calyx is formed by two large plates and one small plate and the arms are pinnulated. This genus is found in Europe and North America in strata of Devonian and Carboniferous age. (K.)

Periechocrinus has a cone-shaped calyx which grows up to 6 cm in diameter and has three basal plates; the arms are branched and the stem circular in section. This genus is found in Europe in strata of middle Silurian age. (K)

Woodocrinus has a small calyx set on a stem of circular ossicles. The stout arms branch into four after a short distance. The short stem tapers to a point and is without roots. This genus is from the Carboniferous strata of Europe. (CP)

Scyphocrinites is a genus with a large calyx and many root branches. It is held to the sea bed by a bulbous 'hold-fast'. This picture shows detail of a calyx and arms and is from the Devonian strata of north Africa. The genus ranges from upper Silurian to Devonian age and occurs in Europe, North America, north Africa and Asia. (C)

Macrocrinus has a small calyx with straight or concave sides and between twelve and twenty stout arms, here showing their pinnules. It is a well-known genus from the lower Carboniferous strata of the USA. (C)

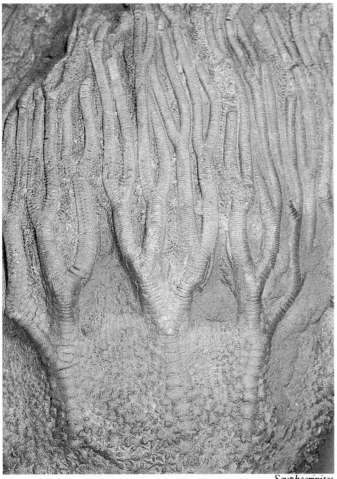

Scyphocrinites

Macrocrinus

Climacograptus and *Diplograptus*

Glyptograptus

Dictyonema

Dictyonema

Phyllograptus

Phyllograptus

Monograptus

Monograptus

Didymograptus

Didymograptus

Monograptus

Graptolitic black shales

Phylum Hemichordata, Class Graptolithina

Graptolites are extinct marine, colonial organisms. Their basic structure is simple. The whole skeleton (the rhabdosome) has a branch (or branches) called a stipe bearing small cups (thecae) in which lived a tiny zooid. The thecae vary in shape and position in different species. Those with thecae on both sides of the stipe are biserial, while if the thecae are only on one side of the stipe the term uniserial is used. The whole rhabdosome is usually only a few centimetres long. The earliest dendroid graptolites (eg *Dictyonema*) had many very small interjoined stipes forming a net-like structure which was cone-shaped in life, but when flattened on a bedding plane of rock becomes triangular. These are thought to have been benthonic. In the early Ordovician period planktonic forms developed. These had very few stipes and larger thecae. Graptolites can sometimes be found in great numbers, and they evolved rapidly to produce a great variety of forms. For these reasons they are used as zone fossils, especially for the Silurian period, though because of their delicate structure they are only found fossilized in very fine sediment such as mudstone and shale. Usually graptolites are left as carbonaceous or pyrite impressions on the rock surface. Three-dimensional material has rarely been recovered from limestones or other strata and such material has allowed detailed microscopic study.

Climacograptus A biserial form with sigmoidal thecae which have upward facing apertures. This genus was probably planktonic, with a vertical position in the water. Its distribution is in lower Ordovician to lower Silurian strata, worldwide. (K)

Diplograptus A biserial genus with sigmoidal thecae which tend to overlap each other. A planktonic and vertical mode of life is probable for this form, which lived from the lower Ordovician to lower Silurian period and was found worldwide. (K)

Dictyonema ranges in size from 2 to 25 cm, and has a conical, net-like rhabdosome with a triangular outline. The many stipes are joined by transverse dissepiments. Of worldwide distribution, this genus was probably benthonic and lived from upper Cambrian to Carboniferous times. (K)

Monograptus has a uniserial rhabdosome between 3 and 75 cm in length. The thecae vary greatly between the many species of this genus, from simple to hook-shaped and sigmoidal. The stipe can be straight, curved, or coiled. *Monograptus* lived from the lower Silurian to the lower Devonian period and was worldwide in distribution. (CP, K)

Didymograptus has a characteristic two-stiped 'tuning-fork' rhabdosome. It is uniserial with simple, tube-shaped thecae only found on one side of the stipes. Ranging in size from 2 to 60 cm, *Didymograptus* lived a planktonic existence during the Ordovician period and was distributed worldwide. (K)

Monograptus

Glyptograptus is a biserial form with curved thecae which become straighter towards the end of the rhabdosome. With a vertical planktonic mode of life, this genus had a worldwide distribution from the lower Ordovician to lower Silurian periods. (K)

Phyllograptus has four wide 'leaf-like' stipes, which are arranged to give a quadriserial rhabdosome and simple tubed thecae. Distributed worldwide, this genus was planktonic and is found in Ordovician strata. (K)

Monograptus This detailed photograph shows the structure of the graptolite. The genus has a single stipe with the pointed thecae, looking like the teeth of a saw, only on one side. This is, therefore, a uniserial graptolite. The specimen is preserved in iron pyrites giving it an almost three-dimensional appearance. (CP)

Graptolitic black shales Graptolites, because of their delicate structure, are mainly preserved in very fine grained shales and mudstones, as in this coastal locality of Ordovician age in South Wales.

Euestheria

Glyphea

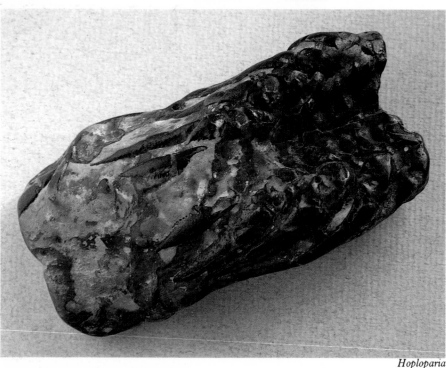

Hoploparia

Phylum Arthropoda

The phylum Arthropoda is characterized by creatures which have a segmented, articulating exoskeleton, which, to allow growth, is periodically moulted (ecdysis). The exoskeleton is composed of chitin with proteins and a strengthening of calcite or calcium phosphate. The symmetry is bilateral and commonly there are paired segmented appendages used for locomotion or other purposes. Arthropods have a well-developed nervous system, the brain being linked to the segments by nerve cords. Blood is circulated by a heart and there are either compound or simple eyes. Some groups have gills for respiration while others breathe through the body surface. Today, arthropods are the most numerous invertebrates and are adapted to all manner of environments from marine to fresh water, terrestrial to aerial. They live in hot, dry conditions and in the Arctic wastes. The fossil record of arthropods begins in the Cambrian period, the most important subphylum being the Trilobita. The other subphyla, Crustacea, including Branchiopods and Xanthoidea (crabs), and Chelicerata, including the Eurypterids, Chilopods and Limulida, are not uncommon in some parts of the fossil record. The phylum includes such creatures as insects, spiders, centipedes, lobsters, crabs, ostracods and scorpions.

Glyphea is a small, shrimp-like decapod with five pairs of limbs used for locomotion. These are invariably missing in fossils. The body, seen in this specimen, is protected by a carapace with a granulose surface; here the head is to the right. At 4.5 cm in length, this example is typical. Their association with *Thalassinoides* trace fossils has led to the suggestion that they produced these burrows (see section on trace fossils page 170). The specimen is from the upper Jurassic strata of northern England; they range from upper Triassic to Cretaceous in Europe, Greenland, east Africa, Australia and North America. (L)

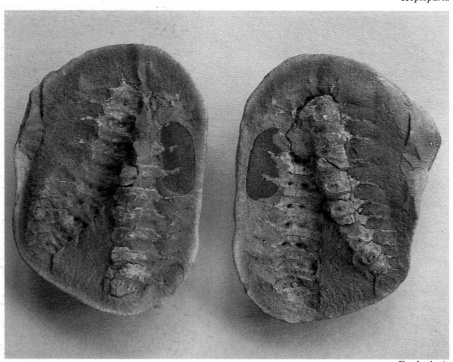

Euphorberia

Euproops

Archaeogeryon

Pterygotus

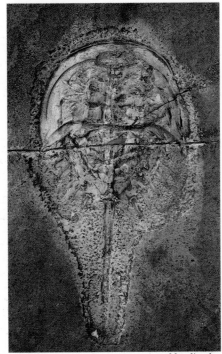

Pterygotus

Mesolimulus

Euestheria belongs to the class Branchiopoda in the sub-phylum Crustacea. The Branchiopods have varied morphological characteristics, this genus having a bivalved carapace which resembles that of a small mollusc, even showing 'growth lines' left by moulting. The genus grows to about 1 cm and is found worldwide in Triassic and Jurassic strata. (K)

Hoploparia A lobster from the Eocene strata of southern England. The segmented legs can be clearly seen. As with crabs, lobsters are frequently preserved in clay nodules. This specimen is 6 cm long. (JHF)

Euphorberia is a Chilopod related to modern millipedes and centipedes. The specimen, which is 3 cm long, has been preserved in an ironstone nodule, both halves of which are shown. The elongate body and slender appendages can be seen. The genus is found in the upper Carboniferous strata of North America and Europe. (S)

Archaeogeryon is a fossil crab from the Miocene strata of South America. Fossils of these arthropods are not uncommon in certain strata and are often preserved in concretions in beds of clay. (L)

Euproops is related to the modern horseshoe crab. The abdomen has raised pleural ridges which cross the flat rim and then become short spines. This specimen shows the dorsal surface with the genal spines extending about halfway down the length of the thorax. It is from the upper Carboniferous strata of Staffordshire, UK, is about 4 cm long and is preserved in a nodule. The genus is found in upper Carboniferous and Permian strata in North America and Europe. (S)

Pterygotus A genus of Eurypterids which were scorpion-like members of the aquatic Palaeozoic fauna. With a segmented body, large eyes and strong front claws they were fierce predators. These arthropods grew to over 1.5 metres, though fossils are rarely more than the 20 cm of the smaller specimen. They were able to move with paddle-shaped limbs. The larger specimen is from Scotland. (L, RMS)

Mesolimulus A fossil king crab from the Jurassic strata of Germany. This genus is characterized by the large dorsal shield and long pointed tail. Growing to about 25 cm maximum, their average is nearer the 12 cm of this specimen. They occur in marine strata from Devonian to Recent age. (L)

Elrathia

Paradoxides

Olenellus

Angelina

Angelina

Asaphus

Asaphellus

Niobella

Platycalymene

Asaphellus

Illaenus

Eodiscus

Ogygopsis

Phylum Arthropoda, Subphylum Trilobita

This is one of the earliest groups to develop a hard exoskeleton, first appearing in the Cambrian period and dying out during the Permian. Fossil trilobites are used as zone fossils for the Cambrian period.

The complex exoskeleton could be articulated and enrolled, possibly for defensive purposes. The various spines and knobs which some possessed may also have been to deter predators (possibly Nautiloids or other trilobites), and some fossils show evidence of predation and damage. The exoskeleton has three lobes running lengthwise – the central axial lobe and one on each side made of ribs (pleurae). There is a thorax, a cephalon (head-shield) with central glabella and eyes, and a pygidium or tail. The exoskeleton was moulted during growth and it is possible for one trilobite to produce many fossils. Very often fossil material is of an incomplete trilobite, just a head-shield, for example, as the exoskeleton naturally breaks at a number of points. Larval stages only 0.1 or 0.2 cm in size have been found as fossils.

The eyes are one of the fascinating aspects of trilobite morphology. These had compound calcite lenses, often beautifully preserved, and they probably provided good vision; a few groups were blind. Exceptional fossil material has in some cases (eg *Triarthrus*) revealed details of the soft appendages, including legs and antennae. The trilobites lived on or near the sea bed, some being active swimmers, others burrowing into the sediment. Trilobite fossils are found in a variety of sedimentary rocks, especially limestones and shales.

Cambrian and Ordovician Trilobites

Elrathia is about 3 cm in length. The cephalon is larger than the pygidium. There are thirteen thoracic segments and the pygidium has five segments. The glabella has weak furrows and the eyes are near the glabella. This genus is common in Cambrian strata in North America. (K, L)

Paradoxides can grow up to 50 cm and is one of the largest trilobites. The thorax has over fifteen segments and the pleurae end in long spines. The genal spines may be over half the body length. The eyes are large and crescent-shaped. The thorax tapers towards the pygidium which is very small. The specimen illustrated here is from Germany, and the genus is well known from Cambrian sediments in east North America, South America, Europe, Turkey and north Africa. (JHF)

Olenellus is up to 8 cm long with a thorax divided into two parts. There are genal spines and spines from thoracic segments. The cephalon is semi-circular and the eyes are large and crescentric. The glabella is furrowed, and the pygidium is very small. This specimen is from British Columbia, the genus being found in Cambrian strata in North America, Greenland and Scotland. (K)

Asaphus grows up to 8 cm long and has eight thoracic segments. The cephalon is triangular with rounded genal angles. The large pygidium has no border. This specimen is from Oslo, Norway, and the genus is found in Ordovician rocks in north-west Europe. (K)

Angelina grows up to 6 cm long and has fifteen thoracic segments. The genal spines extend almost to the pygidium, which is small with only four segments. These specimens are from North Wales and the genus is found in Ordovician rocks in Europe. (K)

Asaphellus is up to 10 cm long and has a thorax with eight segments. The cephalon and pygidium are the same size, the pygidium having very weak segmentation. The eyes are small. These specimens are from Shropshire, UK, and the genus is to be found in Ordovician rocks in South America, China, England and Wales. (K, L)

Niobella may be over 10 cm long and has eight thoracic segments. There is a large cephalon and pygidium, the latter having good segmentation. The eyes are large. This specimen is from North Wales, and the genus is found in Ordovician strata in Europe and North America. (S)

Illaenus grows to 5 cm and has ten thoracic segments. There are distinct furrows which run parallel to each other from the sides of the axial lobe into the cephalon. The axis in the pygidium is very short. The cephalon and pygidium are large and smooth, without segmentation in the pygidium. The specimen is from Oporto, Portugal, and the genus has a worldwide distribution in Ordovician strata. (K)

Platycalymene can be up to 10 cm long, with eleven to thirteen thoracic segments. The glabella has pairs of furrows, and there are small eyes towards the margin of the cephalon. The axis of the pygidium nearly reaches the margin, the pygidium being smaller than the cephalon. The specimen is from Wales and the genus is found elsewhere in the UK and also in Sweden. (S)

Ogygopsis has many segments and a narrow-bordered pygidium which is larger than the cephalon. The glabella is long and furrowed. The eyes are elongated and short genal spines run from the cephalon in complete specimens. The genus can be up to 8 cm in length, and is found in Cambrian strata in North America. This specimen is 3 cm long and is from the Burgess shale, British Columbia, Canada. (W)

Eodiscus is a blind genus with a narrow glabella which has deep lateral furrows. There are three thoracic segments and the pygidium is the same size as the cephalon. It grows to a maximum of 8 cm, is often found fragmented, and is not uncommon in Cambrian strata in Europe and eastern North America. The specimen is from South Wales. (W)

Ellipsocephalus

Ogigiocarella

Agnostus

Paradoxides and *Peronopsis*

Peronopsis

Ogyginus

Agnostus is a blind genus of very small trilobites which reaches a maximum of 1 cm in length. There are two thoracic segments and the pygidium has three or fewer segments. There is a broad furrow across the glabella, which is the same size as the pygidium. Both of these have a wide border. As in this specimen, which is from Vastergotland, Sweden, *Agnostus* is commonly found as fragmented material. It has a worldwide distribution in deepwater sediments of Cambrian age. (W)

Ogyginus may be as long as 4 cm and has eight thoracic segments. The cephalon and pygidium are the same size and the genal spines are well developed, though often broken off in fossils. There are large eyes. The genus is found in Ordovician strata in Europe. (W)

Ogigiocarella is a genus which can grow to over 8 cm though specimens this large are rare. The cephalon is broad but relatively short and the glabella is constricted halfway along. The pygidium is larger than the cephalon. The thoracic axis is very narrow. Large crescentric eyes occur beside the glabella. The genus is found in Ordovician strata in Europe and South America. The naming of this genus has caused much confusion as it has been given many generic names including *Ogigia* and *Ogigiocaris*. (W)

Peronopsis is a very small genus with only two thoracic segments and a cephalon equal in size to the pygidium. This trilobite is about 0.8 cm long in this specimen. The glabella has two lobes and there is a marked border to the pygidium. The genus is from strata of middle Cambrian age and is found in Europe, Siberia and Montana, USA. (K)

Paradoxides and Peronopsis This photograph shows the extreme difference in size of types of trilobites. The small blind *Peronopsis* is 0.8 cm long while the large spinose *Paradoxides* is over 30 cm long and only part of it is shown. The specimen of *Paradoxides* is from South Wales, and that of *Peronopsis* from Montana, USA. (K)

Solenopleura has fourteen segments in the thorax and a wide pygidium with seven or eight segments. The cephalon curves evenly and has a moderately wide border. The tapering glabella has a deep furrow around its margin. The genus is from middle Cambrian strata, and has been found in Europe, Asia, New Zealand and North America. (W)

Meneviella This specimen is of the cephalon, which makes up only about a quarter of the total length of this genus. It is of interest because of the radiating striations, as seen in this example. There is a narrow tapering glabella with three segments. The genus is found in middle Cambrian strata in Britain, Denmark, Asia and eastern North America. (W)

Ellipsocephalus has twelve thoracic segments and a small wide pygidium. The cephalon is large, with a pronounced glabella. The specimen, from Jince in Czechoslovakia, shows three fossils, one of which is an impression. The genus occurs in the lower Cambrian strata of Europe, north Africa and Australia. (W)

Solenopleura

Meneviella

Parabolinella

Trinucleus

Placoparia

Cryptolithus

Flexicalymene

Placoparia

Triarthrus

Bettonolithus

Trinucleus

Leonaspis

Proteus

Encrinurus

Ordovician and Silurian Trilobites

Parabolinella grows to 5 cm and has twenty thoracic segments. The pygidium is very small, but the cephalon is large with long genal spines. The eyes are near the front of the glabella. The specimen is from Shropshire, UK, and the genus is found in strata of Ordovician age in Europe, North America and South America. (K)

Placoparia is up to 4 cm long and has eleven or twelve thoracic segments. There are four blunt spines on the pygidium. The glabella, which is straight-sided, becomes slightly wider towards the front. The specimens are from Santa Justa in Portugal and include one which is enrolled. The genus is found in Ordovician strata in Europe and north Africa. (K)

Trinucleus is up to 3 cm long with six thoracic segments. There is a very pronounced frontal lobe in the glabella, which is in a large cephalon. The genal spines are very long but often missing in fossil material. The cephalon has a fringe with radiating grooves, and the short pygidium is very wide. This specimen, which contains a number of fragments, is from Wales; the genus is commonly found in strata of Ordovician age in England, Wales and the USSR. (K)

Cryptolithus is a diminutive genus which only reaches about 1.5 cm in length. There are six thoracic segments and the comparatively large cephalon has long genal spines and a narrow glabella which becomes wider at the front. The border to the cephalon has radiating grooves. The pygidium is short and wide. The specimen is from Pennsylvania, USA, and the genus occurs in Ordovician strata in Europe and North America. (K)

Triarthrus can be up to 3 cm in length. It has twelve to sixteen thoracic segments. The semi-circular cephalon has wide borders. The eyes are very small and the glabella has two pairs of deep grooves. The small pygidium is triangular with possibly five segments. This genus is of considerable interest because it has been found fossilized with the soft appendages preserved. The specimen is from New York, USA, and the genus has a worldwide distribution in rocks of Ordovician age. (K)

Flexicalymene can be as much as 10 cm long and has twelve or thirteen thoracic segments. The large cephalon has a deeply grooved glabella which is larger than the triangular pygidium. This is a common genus in the Ordovician and Silurian strata of North America and Europe; the specimen illustrated here is from Cincinnati, Ohio, USA. (K)

Bettonolithus The specimen is of a head-shield of this recently re-named genus. It has a wide border with numerous punctations and there is a large central lobe separated from the lateral ones by deep grooves. The genus is found in middle Ordovician strata in central England. (K)

Proteus grows to about 3 cm and has between eight and ten thoracic segments with pronounced pleural furrows. The furrows on the glabella are weak. The pygidium is semi-circular. It ranges from Ordovician to Carboniferous, with a worldwide distribution. The specimen is from Devonian strata of West Germany. (W)

Leonaspis grows to a maximum length of 2 cm. It is very spinose with both genal and pleural spines. There are eight to ten thoracic segments. It ranges from Silurian to Devonian age in North and South America, Australia, Asia and Europe. (W)

Encrinurus has twelve thoracic segments and can be over 5 cm long. The cephalon is bigger than the pygidium and the glabella has distinctive tuberculation. The eyes are on long stalks. It is found worldwide in strata of Ordovician to Devonian age. (W)

Dalmanites

Silurian, Devonian and Carboniferous Trilobites

Dalmanites can grow to 8 cm but is usually smaller and often fragmented in fossils. There are large eyes and the cephalon is a similar size to the pygidium, which has a short spine on its end. The genal spines wrap close to the thorax. The genus is found in strata from Silurian to Devonian age in Europe, the USSR, North America and Australia. (CP)

Scutellum The specimen shows the pygidium which, unlike many genera of trilobites, has marked lateral ribs, radiating from the point at which it joins the thorax. The specimen, which is from the Devonian strata near Prague, Czechoslovakia, has a small portion of the outer surface missing, showing the fine concentric bands on the underlying surface. The genus has been found worldwide in strata ranging from Silurian to Devonian in age. (W)

Eocyphinium belongs to a group of trilobites often called *Phillipsia*. The cephalon has a sub-triangular outline and there are ten thoracic segments. The surface is coarsely pustulate. This genus is found in lower Carboniferous strata in Europe. (W)

Calymene has twelve thoracic segments and may be as long as 10 cm. The cephalon is larger than the pygidium, which has six segments. The glabella has rounded lobes, and a deep groove runs between the glabella and the border of the cephalon. The eyes are large. The genus is found in strata of Silurian and Devonian age in Europe, North America, South America and Australia. (W)

Trimerus is characterized by very indistinct longitudinal lobes, and a very wide thoracic axis. The cephalon has no borders; the pygidium is large and triangular. This genus is found worldwide in strata of Silurian and Devonian age. (JHF)

Acidaspis This specimen has a rather fragmented cephalon, which, when present, has a tapering glabella with three pairs of lateral lobes. The thorax has ten segments and there are spines on the border of the pygidium which is also tuberculate. A genus from the middle Ordovician to middle Devonian strata of Europe and North America. (W)

Bumastus may be 10 cm long but is usually smaller and fragmented. The very wide axis covers three-quarters of the thorax width. The cephalon and pygidium are the same size and are smooth and rounded, the glabella being poorly defined. *Bumastus* is a genus from Silurian sediments in Europe and North America. (W)

Scutellum

Eocyphinium

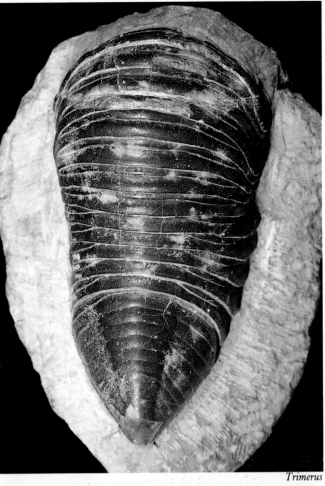

Calymene

Trimerus

Acidaspis

Bumastus

Megalichthys

Paramblypterus

Elonichthys

Elonichthys

Megalichthys

Rhizodopsis

Cephalaspis

Bothriolepis

Fish

Fossils of fish are reasonably common. Living in water, they are rapidly buried by sediment when they die and they have various hard parts which can become well preserved. These include scales, bones (in many genera) and teeth. A brief outline of the evolution of the whole group will help to place those illustrated into the sequence.

The earliest fish are the jawless **Agnathans**, which were in fact the first vertebrates. They had a simple hole for a mouth and many must have grubbed about for organic debris in the bottom sediment of fresh and brackish water. They are known from the Ordovician age and became dominant in Silurian and Devonian times. One group, the **Osterostrachi**, had very flattened heads and a heavy armoured body. The head-shield is continuous apart from two small eye-sockets. They were probably bottom dwellers incapable of good movement. Another group, the **Heterostraci**, had paired nostrils and eyes on the sides of their bony plated heads. The only distant descendants of these ancient fish are the hag-fish and eel-like lampreys.

During the Devonian period the Agnathans died out, to be replaced by the **Gnathostomes**, the jawed fish. There were four main groups which developed at this time, some of which are represented today. The **Acanthodii** have a similar overall shape to modern fish but are spiny and primitive with no dermal bones on the margins of the jaws. The body was covered with many small diamond-shaped scales and the head covering was of small plates. They first appeared in the upper Silurian period and lived through the Carboniferous until the lower Permian period. The **Placoderms** were another important group during the Devonian period. They lived in both marine and freshwater environments and were well protected with a head-shield of heavy plates. The flattened overall shape of many of them suggests that they were bottom dwellers. They had paired limbs and most were quite small, though *Titanichthys* grew to over 9 metres long. These fish were very successful and spread throughout the world in the upper Devonian period. Only a handful survived into the Carboniferous period.

The **Chondrichthyes** are another group first appearing in the Devonian period. These were abundant during the Carboniferous and Permian periods. Only a limited number reached the Jurassic. *Lamna* and *Charcarodon* are well-known genera found fossilized in the Cretaceous and Tertiary strata. This group comprises the cartilaginous fish and included skates, rays and sharks. Typically, the skin is studded with small sharp scales and the mouth is on the underside. They are predatory and many are equipped with sharp, pointed teeth, while others have rounded flatter teeth for crushing prey. The cartilage of their skeletons is not often preserved, though some groups were able to secrete calcium salts in the cartilage. The fossil record of these fish is mainly spines and teeth from marine strata.

The fourth important group of jawed fishes which appeared in the Devonian age are the **Osteichthyes** or bony fish. These are a highly successful group which developed by adaptive evolution into a diverse number of types, many of which are represented today. The **Actinopterygii** are the ray-finned fish. These breathe by gills and evolved gradually through the upper Palaeozoic and into the Mesozoic era, becoming well adapted to catching food by their versatile swimming. Their paired fins can be used for steering, slowing down and increasing power, while the dorsal fin maintains balance. The **Holosteans** have bony plates rather than scales and developed at the start of the Mesozoic era but gave way to the **Teleosts** in the Jurassic period. The Teleosts are the most advanced of the Actinopterygians and are the fishes of today, being represented by some 24,000 species inhabiting all types of aquatic environments.

The **Sarcopterygians** are another group of bony fish, characterized by lobe-fins. These fins have bone and muscle extending into them. They were numerous fish in the Devonian and Carboniferous periods and have two main types, the **Dinopans** and **Crossopterygians**. The former are the lung-fish, of which three genera survive today in tropical freshwater habitats. The Crossopterygians have different skeletons from the lung-fish and were active carnivores, living on smaller fish and invertebrates. Detailed investigation of their fossilized vertebrae shows similarities with those of early amphibians, the labyrinthodonts. During the Devonian period the Crossopterygians developed and became common freshwater fish in the upper Palaeozoic era. They declined in the Mesozoic and only the rather unusual Coelocanth, *Latimeria*, survives at the present time.

Megalichthys is a well-known Crossopterygian genus from the Carboniferous age. It grows to 2 m but specimens are often fragmentary and smaller than this. The scales are rhomboid and it has lobate paired fins; the tail is diphycercal (symmetrical around the vertebral column). The specimens are from the upper Carboniferous strata of Staffordshire, UK. The upper specimen is part of a skeleton, the lower one shows the rhombic scales. (S)

Paramblypterus is a genus of Actinopterygian fish from the upper Carboniferous period. The fins are triangular and the scales rhomboid. The tail is heterocercal with the vertebral column going through the upper lobe. The specimen is from a nodule in the upper Carboniferous strata of West Germany. (S)

Elonichthys ranges from the lower Carboniferous to the upper Permian periods and is an Actinopterygian fish. This specimen from Staffordshire, UK, shows both halves of a nodule containing the fossil. The scales are rhomboid and the tail, in a complete specimen, is deeply cleft and heterocercal. The fins are triangular. (S)

Rhizodopsis is a genus of Crossopterygian fish of Carboniferous age. The body is slender with thick rhomboid scales, well shown in this specimen from Staffordshire, UK. The paired fins are lobate and the tail heterocercal. This genus inhabited shallow water. (S)

Cephalaspis is a reasonably common Agnathan fish which is jawless and has no paired fins. There is a massive head-shield with two small eye holes placed centrally on top. The small mouth, probably adapted for suction feeding, is on the underside, with pairs of gill slits behind it. The Agnathans are the simplest of fish with an eel-like body. They inhabited fresh and brackish water, and this genus lived from the upper Silurian to middle Devonian period. The specimen is from Angus, Scotland. (RMS)

Bothriolepis is a genus of Placoderm fish which have paired fins and primitive jaws. This particular genus, from the upper Devonian rocks of Scaumenac Bay in Canada, has protective shields on the head and anterior part of the body. There are two mobile bony appendages attached to the body shield. The rest of the body is virtually naked, being without scales. The tail is heterocercal and there is no anal fin. The picture shows the head and body shields. The genus lived in fresh water and was probably a mud swallower. It has a worldwide distribution occurring from North America to Antarctica. (RMS)

Ischnacanthus

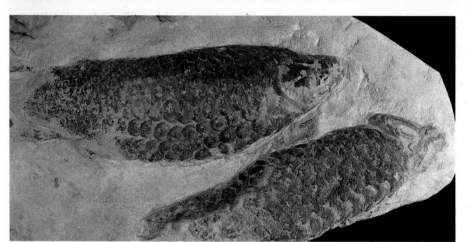

Holoptychius

Gyroptychius

Ischnacanthus is one of the Acanthodians, the first true jawed fish. It has numerous small teeth and the body is characterized by two dorsal and three ventral spines, which are easily seen in this specimen from the lower Devonian strata of Kinnaird, Scotland. (RMS)

Holoptychius is a Crossopterygian genus ranging from the upper Devonian to lower Carboniferous period. It has a heterocercal tail and lobed fins with dermal bones. The scales are large and cycloid. This genus grew to a maximum length of about 50 cm. The specimens are from the classic locality at Dura Den in Fife, Scotland. (RMS)

Gyroptychius is a middle Devonian genus of Crossopterygian fish. It has a slender body and a diphycercal tail which is symmetrical around the vertebral column. There are two dorsal fins, plus pectoral, pelvic and anal fins. The specimen is from Orkney, Scotland. (RMS)

Glyptolepis is a Crossopterygian fish from the Devonian period. The scales are thin and rounded with small tubercles or ridges. This specimen, from Nairn, Scotland, is preserved in a concretion, with good preservation of the scales. The genus swam with an eel-like movement of its elongate body. (D)

Cheirolepis is an Actinopterygian genus from the middle and upper Devonian period which grew to a length of about 35 cm. It has slightly lobate pectoral fins and the body is covered with minute square scales. There is a single dorsal fin. The specimen is from Clune, Scotland. (RMS)

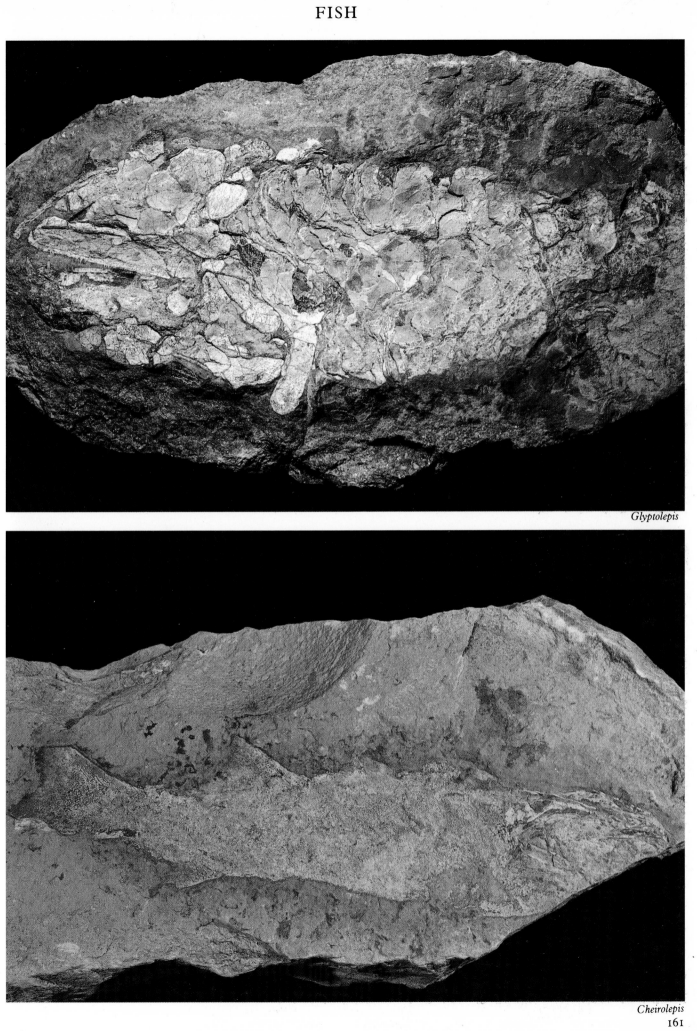

Glyptolepis

Cheirolepis

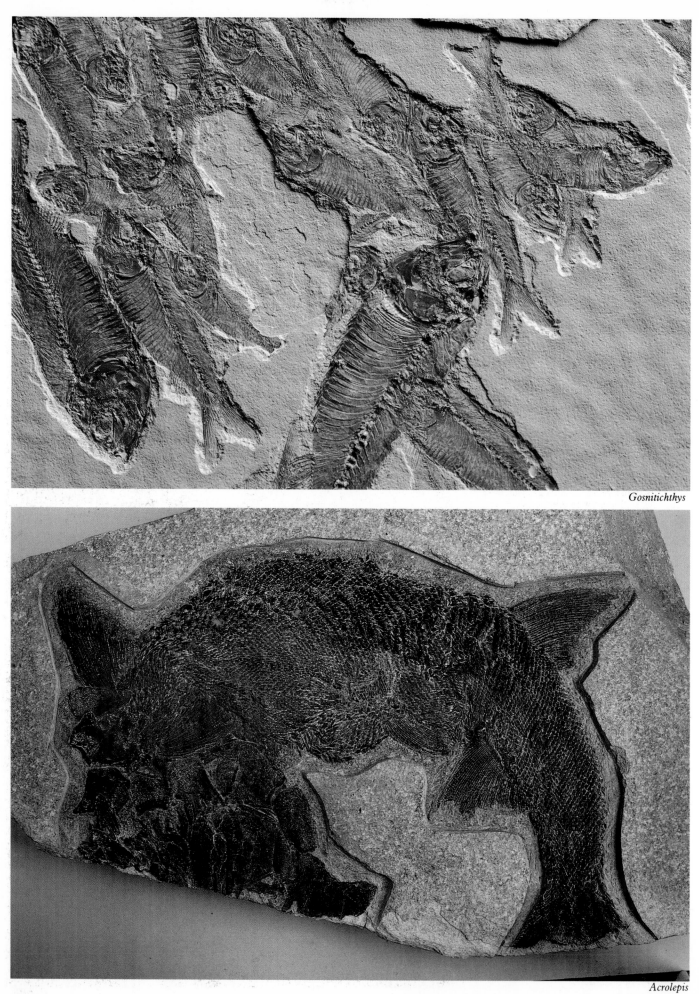

Gosnitichthys

Acrolepis

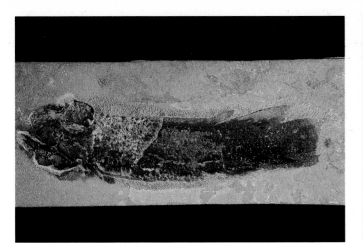

Dipterus

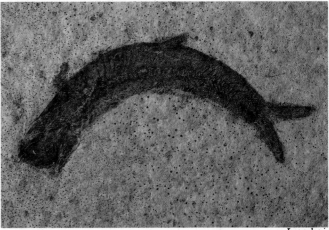

Leptolepis

Dipterus is a Dinopan lung-fish from the middle Devonian period, and is a medium-sized, long-lived group characterized by a heterocercal tail and bony strengthening of the fins. The scales are thin, cycloid and overlapping. There are typical lung-fish grinding tooth-plates in both jaws. The specimen is from the Achnaharras fish beds in Scotland. (RMS)

Platysomus is a genus of Actinopterygian fish ranging from lower Carboniferous to upper Permian age. It has a very deep body, elongate unpaired fins and the scales are dorsoventrally elongated. The teeth are pointed and conical. The specimen is from the upper Permian strata of Durham, UK. (C)

Dapedius is a Holostean fish from the lower Jurassic period. This specimen from southern England shows the characteristic rounded shape with a long dorsal fin. The scales are rectangular and the head covered with larger plates. The small mouth is full of sharp, thin teeth. (L)

Gosnitichthys belongs to the modern Teleost group of fish. It was a surface feeder which grew to a length of about 50 cm. Fossils of this genus are often very numerous on bedding planes, as here in a specimen from the famous Green River Formation of Wyoming, USA. This deposit is of Eocene age and these fish were fossilized when their environment dried rapidly. (CP)

Leptolepis is a member of the Osteichthyes (the bony fish) and is very similar to modern fish. This genus is commonly small, with a tapering body. The small head is armed with sharp teeth. Many fossils of this genus are found in sediments from upper Triassic to upper Cretaceous age throughout Europe, the USA, South Africa and Asia. (C)

Acrolepis is commonly found as large specimens, this one from the Permian strata of Durham, UK, being about 40 cm long. The head is about a quarter of the whole length and the pelvic fins are half the size of the pectoral fins. The body is covered with very coarse scales. It is found in Carboniferous to upper Permian strata in Britain, Germany, Africa, the USSR and Greenland. (C)

Platysomus

Dapedius

Charcarodon

Charcarodon

Charcarodon

Lamna

Charcarodon

Lamna

Oxyrhina

Rhizodus

Oxyrhina

Megalichthys

Odontaspis

Odontaspis

Ptychodus

Asterocanthus

Pliosaur

Ichthyosaur

Ceratodus

Teeth

Charcarodon   Only the teeth of this large shark are commonly found. These triangular teeth have serrated cutting edges. The shark attained a maximum length of about 15 m, and is found worldwide in strata of Tertiary age. (K, L)

Lamna   A medium to large shark with teeth which lacked serrated edges, but which have smaller side spikes; the right-hand one is visible through the sediment which encloses the specimen. It has a worldwide distribution in Cretaceous to Pliocene strata. (K, L)

Oxyrhina is a member of the Lamnidae (sharks); its teeth differ from *Lamna* in the lack of side denticles. It occurs in Cretaceous and Tertiary strata in Europe, India, the USA and China. (L)

Rhizodus is a Crossopterygian fish which grew to large size. The lower jaw may be a metre in length, and the teeth up to 20 cm. The specimen is from the Carboniferous Scottish oil shales. The genus is found worldwide in strata of upper Devonian and lower Carboniferous age. (K)

Megalichthys is a genus of Carboniferous age which is illustrated elsewhere. (S)

Ptychodus   This well-known flattened tooth is from a shark which is found in Cretaceous strata in Europe, Asia, Africa and North America. The parallel oval ridges are ideal for crushing hard prey such as mollusc shells. (L)

Odontaspis   This illustrates the long sharp teeth from a genus of sharks which is still living today. The teeth have small side cusps. The

modern representatives of the genus reach about 4 m in length. The fossil forms are found in lower Cretaceous to Recent strata in North and South America, Asia, Africa, New Zealand and Europe. (L)

Asterocanthus   This rounded elongate tooth is from a shark genus found in strata of upper Triassic to Palaeocene age in Europe, the Middle East, north Africa and North America. Such teeth were used for grinding prey with hard shells. (K)

Pliosaur   A tooth from a middle Jurassic marine reptile of a type which developed from the Plesiosaurs and is now classified with that group. The Pliosaurs are characterized by a

short thick neck and a very large head up to 3 m long, the body being up to 12 m in length. (K)

Ichthyosaur   A detailed picture of the cone-shaped, striated teeth of this marine reptile. Fragments of the jaw are not uncommon as fossils in Jurassic sediments. The teeth in this example from South Wales are 2 cm long. (W)

Ceratodus   The most frequently found fossils of this dinopan fish from the Mesozoic era are its teeth. As seen here, these are fused into plates and used for crushing such items as shellfish. The specimen, which is 2 cm long, is from the Triassic strata of southern England. This lung-fish is very closely allied to the modern Australian form *Neoceratodus*. (JM)

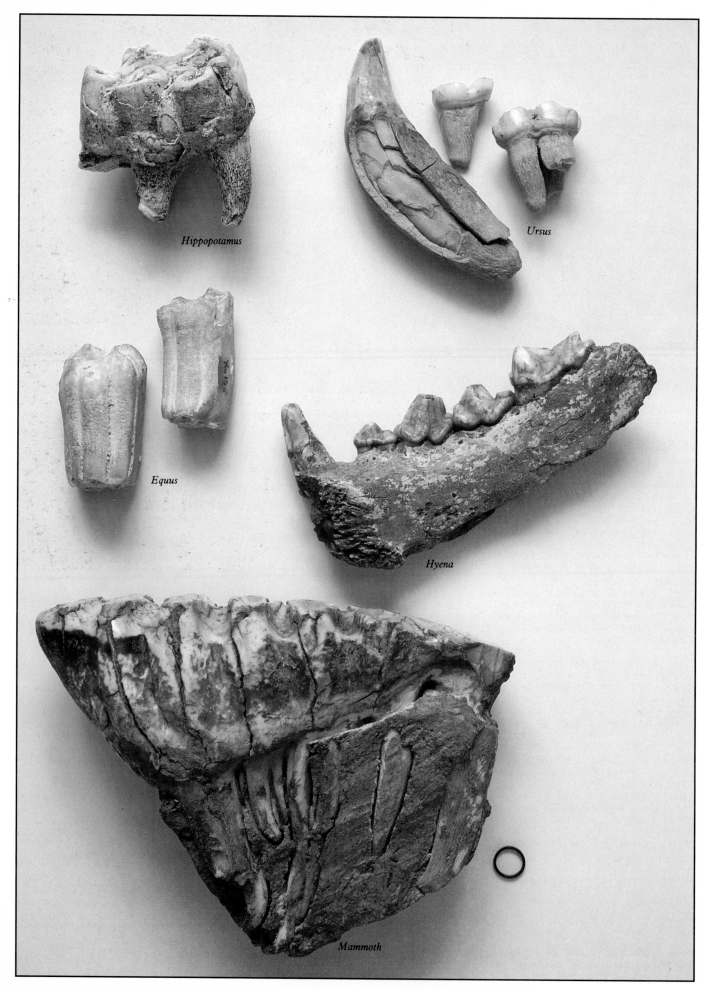

Hippopotamus

Ursus

Equus

Hyena

Mammoth

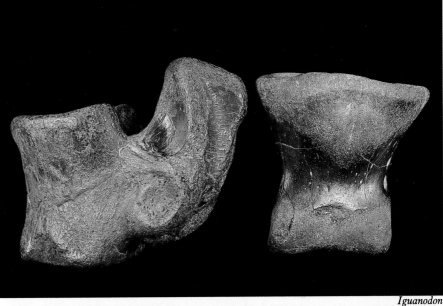

Iguanodon

Temnospondyl

Other Vertebrates

Fossils of vertebrates other than fish, i.e. amphibians, reptiles, birds and mammals, are not common. Usually, they are fragmented fossils consisting of individual bones or teeth. Land vertebrates are even less common as fossils than the remains of those that lived in the sea, but in a few regions, for example parts of Africa, Asia and North America, there are huge numbers of fossilized land-vertebrate skeletons.

Dinosaurs are among the best-known land-dwelling extinct vertebrates and there is good fossil evidence of this popular group. They are found in river and lake sediments of Jurassic and Cretaceous age in America, China, Central Asia, Australia, Africa and Europe, often in remote regions, and it is very possible for the amateur to make exciting finds.

Dinosaurs developed from a more primitive group of reptiles called the Thecodonts which themselves developed during the Carboniferous period. Crocodiles also developed from the Thecodonts. Three main types of dinosaur recognized are: the Sauropodomorphs, which are the large herbivorous creatures like *Diplodocus*; the Theropods, a group of bipedal carnivores, commonly with much reduced front limbs, as in *Tyrannosaurus*, and the Ornithischia which are both bipedal and quadruped and lived in herds like modern herbivorous mammals. *Iguanadon* is a large example, which grew to 10 m and possessed a large thumb spike possibly used for defence or for dragging branches towards its mouth. Two very specialized groups are the flying Pterosaurs, the exact relationships of which are uncertain, and the marine reptiles – like the Plesiosaurs and Ichthyosaurs – which have a good fossil record.

Mammal fossils appear in the geological record later than those of reptiles, first appearing in strata of Triassic age. Early mammals appear to have been small, shrew-like creatures and their remains are often in the form of skull fragments. Up until the Cretaceous period mammal fossils are meagre, but in the Tertiary strata they become more common. Some deposits, such as those of Oligocene age in South Dakota, USA, contain abundant mammal fossils, but, as with other land-dwelling groups, identification is often difficult as it is based on fragments of bone and the structure of teeth. Pleistocene river gravels are often productive deposits in which to search for mammal fossils like mammoth teeth, whilst whale ear bones occur in marine strata from Miocene to Recent age.

Land Vertebrates

Hippopotamus This specimen is of canine teeth from the upper jaw. The lower canines are far larger. This genus appeared in the upper

Pliocene in Asia and Africa and spread through Europe during Pleistocene interglacials. Fragments such as these are not uncommon in river gravels deposited at this time. (S)

Ursus These teeth are from the cave bear, and include incisors, large canine fangs and molar teeth. Such fossils are often encountered in cave deposits. These specimens are from Pleistocene strata in a limestone cave in Devon, UK. (K)

Equus Molar teeth from the horse genus have square crowns with a very complex pattern on the upper surface. The genus includes not only horses but zebras and donkeys. Teeth such as this can be found in late Tertiary and Recent deposits in Europe, Africa, Asia, North and South America. (K)

Hyena The lower jaw of a cave hyena from Pleistocene deposits. This has the dentition of a carnivore adapted for crushing bones. From the upper Miocene to Recent strata of Europe, Africa and Asia. (K)

Mammoth Here is a massive specimen which is a fragment of the cheek teeth. These consist of many joined segments making ridges on the

upper surface which are efficient chewing mechanisms. These large fossils occur in Pleistocene deposits, often river gravels, in Europe, Africa, Asia and North America. (S)

Iguanodon These specimens, which come from Sussex, UK, are of a caudal vertebra (*left*) and a toe bone. This dinosaur reached a height of 10 m and had a bipedal stature. It was a common genus in the Cretaceous period of Europe and probably lived in herds, as skeletons have been found in considerable numbers in Belgium and West Germany. (HST)

Temnospondyl This is a very rare fossil; in fact, it is the earliest known complete amphibian skeleton. Such finds are often the reward of diligent painstaking searching, by amateurs as well as professional palaeontologists. This specimen, which is about 40 cm long, and is from the lower Carboniferous strata of Lothian, Scotland, is contained on a slab of limestone and shows the skull, backbone and limbs, including the delicate feet. The specimen is also of interest because it shows no indication of a completely aquatic life-style. (RMS)

Balaena

Myliobatus

Ichthyosaur

Argillochelys

Ichthyosaur

Ichthyosaur

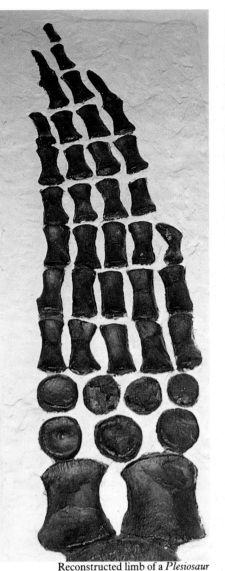

Reconstructed limb of a *Plesiosaur*

Plesiosaur vertebra

Plesiosaur ribs

Marine Vertebrates

Balaena The curved ear-bones of whales are not uncommon in strata of Miocene and younger age. This and related genera are entirely marine and have many fish-like characteristics, not unlike those of the Mesozoic marine reptiles. The earliest whales appeared in the Eocene period, and the modern types, which are among the most intelligent mammals, had developed by the Miocene period. (S)

Myliobatus A genus of ray-fish; this specimen is from the Eocene strata of southern England. The item illustrated is part of the lower tooth plate which are the parts of the fish most commonly found as fossils. These plates were arranged in rows for grinding food. The characteristic grooved surface can be seen clearly. Modern Myliobates live in relatively warm waters and feed on molluscs, the shells of which are easily smashed by the powerful dentition. Fossils of this genus have been found in New Zealand, Asia, Europe, Africa and North America. (HST)

Ichthyosaur Three fragments of this wonderfully streamlined Mesozoic reptile are shown. The large, rounded slab contains a sectioned skull, to the right of which is a fragment of one of the creature's paddle-like limbs. At the bottom of the page is a jaw showing the carnivore's teeth. Specimens such as these are common in marine strata of Jurassic age, as are vertebrae and other bones. Occasionally whole skeletons are recovered and examples are known containing fossilized young. One such deposit is that of lower Jurassic age at Holzmaden in Germany. (K)

Argillochelys This specimen is the femur of a marine turtle from strata of Eocene age in southern England. It first appeared in upper Jurassic times in Europe and by the lower Cretaceous had reached North America and Asia. In strata of Miocene age it occurs in Africa. Very often fossils are of sections of the carapace, which are called scutes. (HST)

Plesiosaur These were quite large marine reptiles which grew to over 10 m in length, though usually specimens are nearer 3 m. They are not as 'fish-like' in their appearance as the ichthyosaurs, but have a broad body with strong, close ribs, as in this example from the lower Jurassic strata of North Yorkshire, UK. (Wh)

Plesiosaur vertebra Often only single bones or a short string of vertebrae are found, the whole skeleton having become disarticulated after death, before fossilization. This vertebra is about 10 cm in diameter. (D)

Reconstructed limb of a Plesiosaur from the lower Jurassic strata of North Yorkshire, UK, shows the 'paddle-shape' required for effective propulsion. These creatures moved by flapping their limbs up and down rather like the action a modern penguin uses. Plesiosaurs are common in rocks of Jurassic and Cretaceous age in many parts of the world. (Wh)

Imbrichnus

Cruziana

Serpula

Rotularia

Lithophaga

Skolithus

Scolicia

Satapliasaurus

Thalassinoides

Skolithus

Trace Fossils

Sedimentary strata contain a variety of structures, some of which are inorganic in origin, but many shapes and forms which occur in and on beds of rock are of organic origin. These trace fossils can be defined as sedimentary structures produced by biological activity. They include tracks and trails, footprints, burrows, marks made by feeding activity and resting hollows. A variety of creatures is responsible for leaving this evidence of their activities, including worms, molluscs, arthropods and vertebrates.

Trace fossils are of great use in palaeoecology, the reconstruction of past habitats, especially when we can compare a burrow or trail with a similar modern one. These tracks, trails and footprints are also valuable as 'way-up criteria', which allow geologists to work out the original orientation of a stratum. This is of considerable value when the principle of superposition is being applied in stratigraphic work. If beds have been inverted by, for example, recumbent folding, burrows will go up from bedding surfaces and footprints would appear on the under side of a stratum.

Trace fossils are usually preserved as casts or moulds; the cast is a replica made by the infilling (with sediment) of the burrow, which is a mould. These fossils, like shells, bones and other obviously biological material, are given scientific names. The name of a burrow will have a different name from the animal that made it, and often it is not known which animal made the trace fossil.

Worm burrows are common trace fossils. The worms themselves, because of their soft bodies, are extremely rare as fossils. The burrows are usually vertical tubes running down from a sediment surface, which are often infilled with sediment of a slightly different colour from that of the bed itself. Some burrows are U-shaped and may be confused with the mining activities of amphipods. The exact type of worm that constructed a given burrow is often unknown, though some types like serpulids (which leave calcareous tubes) and terebellids (whose tubes are cemented with shell fragments) are more obvious. Arthropods such as trilobites and shrimps leave various signs of their movements. Many trilobites were deposit feeders that rummaged through the detritus on the sea floor.

As they did this they created trails which have a characteristic pattern. *Cruziana* are linear trilobite trails which have scrapes at an oblique angle. *Rusophycus* is a slight excavation in the sediment which could have been a resting place. Shrimps make burrows which are vertical, but have horizontal passages at the base. Bivalve molluscs like *Mya* burrow deep into sediment, as does the Brachiopod *Lingula*. These burrows often contain the fossilized shell of the animal. Gastropods graze on the sea bed and their wandering trails are frequently found on bedding planes. Possibly the best known trace fossils found in non-marine sediments are the footprints of large reptiles, and coprolites (fossil excrement). The branch of palaeontology called 'ichnology' is devoted to the study of footprints. From the footprints and trackways found as trace fossils it is possible to determine whether a two- or four-footed creature made the tracks, how fast it moved and what it weighed.

Imbrichnus is a sediment-filled burrow, winding parallel to the bedding plane, 0.5 to 1 cm in diameter, and possibly made by a small mollusc. The specimen is from the upper Jurassic period of southern England. (K)

Cruziana are feeding trails of trilobites. This example is from the Ordovician strata of Rennes, Brittany, France, and shows the characteristic grooved pattern. (K)

Rotularia are the spiral tubes of polychaete worms, made of calcareous secretions. This example is from the Eocene strata of Sussex, UK. (K)

Serpula are worm tubes, often found on the shells of large molluscs. This specimen is from the Cretaceous strata of Hastings, UK. (K)

Rounded burrows of Lithophaga, a bivalve mollusc, and thin burrows of polychaete worms. A specimen from the Mendip Hills, Somerset, UK. (K)

Skolithus This specimen is a section through worm burrows in quartzite of Cambrian age from Sutherland, UK. (CP)

Scolicia The large curving structures are the grazing trails of gastropods. The specimen is of Carboniferous sandstone from West Yorkshire, UK. (K)

Satapliasaurus The footprint of a bipedal dinosaur from the middle Jurassic period of Yorkshire, UK. The impression was left in wet sand as the creature moved across marshland. This specimen is the positive cast of the footprint made by sediment which filled in the original print. Prints from this area vary between 10 and 60 cm in length; this one is 15 cm long. (JM)

Thalassinoides This bedding plane, of middle Jurassic age, is covered with the branching and Y-shaped burrows probably made by a decapod crustacean. Such burrows are made in the soft sediment just below the sea bed. The crustacean *Glyphea* figured with the arthropods has been found fossilized in burrows of this type in sediments of sub-tidal origin. (C)

Skolithus This cliff section, of which about 2 m is shown here, is of Cambrian quartzite with vertical worm burrows. These burrows are shown in section on the larger plate. The sediment was probably formed in inter-tidal mud-flat conditions.

Further Reading

Arduini, P., Teruzzi, G., *The Macdonald Encyclopedia of Fossils* (Macdonald, London and Sydney, 1986)

British Museum (Natural History), *British Caenozoic Fossils* (London, 1982)

British Museum (Natural History), *British Mesozoic Fossils* (London, 1982)

British Museum (Natural History), *British Palaeozoic Fossils* (London, 1982)

Clarkson, E. K. M., *Invertebrate Palaeontology and Evolution* (Allen and Unwin, London, 1979)

Dana, J. D., *Textbook of Mineralogy* (Wiley and Sons, New York, 1932)

Deer, W. A., Howie, R. A., Zussman, Z., *An Introduction to Rock-forming Minerals* (Longmans, London, 1966)

Desutels, P. E., *The Mineral Kingdom* (Hamlyn, London, 1969)

Francis, P., *Volcanoes* (Penguin, London, 1976)

Gass, I. G., Smith, P. J., Wilson, R. C. L., *Understanding the Earth* (Artemis, Horsham, Sussex, 1971)

Geological Museum, London, *Volcanoes* (HMSO, London, 1977)

Hallam, A., *Planet Earth* (Elsevier Phaidon, Oxford, 1977)

Halstead, L. B., *Evolution of Mammals* (Lowe, London, 1978)

Halstead, L. B., *Hunting the Past* (Hamish Hamilton, London, 1982)

Hamilton, W. R., Wooley, A. R., Bishop, A. C., *The Country Life Guide to Minerals, Rocks and Fossils* (Country Life, London, 1983)

Hatch, F. H., Wells, A. K., Wells, M. K., *Petrology of the Igneous Rocks* (Murby, London, 1961)

Holmes, A., *Principles of Physical Geology* (Nelson, London and Edinburgh, 1966)

Kirkaldy, J. F., *Fossils in Colour* (Blandford, London, 1976)

Kirkaldy, J. F., *Minerals and Rocks in Colour* (Blandford, London, 1976)

Miles, R. S., *Palaeozoic Fishes* (Chapman Hall, London, 1971)

Mondadori, A., *The Macdonald Encyclopedia of Rocks and Minerals* (Macdonald, London, 1986)

Moore, R. C. (ed), *Treatise on Invertebrate Palaeontology* (University of Kansas, Kansas, 1953 onwards)

Murray, J. W., *Atlas of Invertebrate Macrofossils* (Longman, Harlow, Essex, 1985)

Norman, D., *Illustrated Encyclopedia of Dinosaurs* (Salamander, London, 1985)

Pellant, C., *Earthscope* (Salamander, London, 1985)

Pettijohn, F. J., *Sedimentary Rocks* (Harper and Brothers, New York, 1949)

Read, H. H., Watson, J., *Introduction to Geology* (Macmillan, London, 1974)

Romer, A. S., *Vertebrate Palaeontology* (University of Chicago Press, Chicago, 1966)

Rutley, F., *Elements of Mineralogy*, Ed Read, H. H. (Allen and Unwin, London)

Thackray, J., *British Fossils* (HMSO, London, 1984)

Turner, F. J., Verhoogen, J., *Igneous and Metamorphic Petrology* (McGraw-Hill, New York, 1960)

Watson, J., *Geology and Man* (Allen and Unwin, London, 1983)

Whitten, D. G. A., Brooks, J. R. V., *The Penguin Dictionary of Geology* (Penguin, London, 1972)

Rocks Index

Minerals Index

Fossils Index

INDEX